T0212408

Springer Complexity

Springer Complexity is an interdisciplinary program publishing the best research and academic-level teaching on both fundamental and applied aspects of complex systems—cutting across all traditional disciplines of the natural and life sciences, engineering, economics, medicine, neuroscience, social and computer science.

Complex Systems are systems that comprise many interacting parts with the ability to generate a new quality of macroscopic collective behavior the manifestations of which are the spontaneous formation of distinctive temporal, spatial or functional structures. Models of such systems can be successfully mapped onto quite diverse "real-life" situations like the climate, the coherent emission of light from lasers, chemical reaction-diffusion systems, biological cellular networks, the dynamics of stock markets and of the internet, earthquake statistics and prediction, freeway traffic, the human brain, or the formation of opinions in social systems, to name just some of the popular applications.

Although their scope and methodologies overlap somewhat, one can distinguish the following main concepts and tools: self-organization, nonlinear dynamics, synergetics, turbulence, dynamical systems, catastrophes, instabilities, stochastic processes, chaos, graphs and networks, cellular automata, adaptive systems, genetic algorithms and computational intelligence.

The three major book publication platforms of the Springer Complexity program are the monograph series "Understanding Complex Systems" focusing on the various applications of complexity, the "Springer Series in Synergetics", which is devoted to the quantitative theoretical and methodological foundations, and the "Springer Briefs in Complexity" which are concise and topical working reports, case studies, surveys, essays and lecture notes of relevance to the field. In addition to the books in these two core series, the program also incorporates individual titles ranging from textbooks to major reference works.

Editorial and Programme Advisory Board

Henry D. I. Abarbanel, Institute for Nonlinear Science, University of California, San Diego, USA

Dan Braha, New England Complex Systems Institute and University of Massachusetts Dartmouth, USA

Péter Érdi, Center for Complex Systems Studies, Kalamazoo College, USA and Hungarian Academy of Sciences, Budapest, Hungary

Karl J. Friston, Institute of Cognitive Neuroscience, University College London, London, UK

Hermann Haken, Center of Synergetics, University of Stuttgart, Stuttgart, Germany

Viktor Jirsa, Centre National de la Recherche Scientifique (CNRS), Université de la Méditerranée, Marseille, France

Janusz Kacprzyk, System Research, Polish Academy of Sciences, Warsaw, Poland

Kunihiko Kaneko, Research Center for Complex Systems Biology, The University of Tokyo, Tokyo, Japan

Scott Kelso, Center for Complex Systems and Brain Sciences, Florida Atlantic University, Boca Raton, USA

Markus Kirkilionis, Mathematics Institute and Centre for Complex Systems, University of Warwick, Coventry, UK

Jürgen Kurths, Nonlinear Dynamics Group, University of Potsdam, Potsdam, Germany

Ronaldo Menezes, Florida Institute of Technology, Computer Science Department, 150 W. University Blvd, Melbourne, FL 32901, USA

Andrzej Nowak, Department of Psychology, Warsaw University, Poland

Hassan Qudrat-Ullah, School of Administrative Studies, York University, Toronto, ON, Canada

Linda Reichl, Center for Complex Quantum Systems, University of Texas, Austin, USA

Peter Schuster, Theoretical Chemistry and Structural Biology, University of Vienna, Vienna, Austria

Frank Schweitzer, System Design, ETH Zürich, Zürich, Switzerland

Didier Sornette, Entrepreneurial Risk, ETH Zürich, Zürich, Switzerland

Stefan Thurner, Section for Science of Complex Systems, Medical University of Vienna, Vienna, Austria

Understanding Complex Systems

Founding Editor: S. Kelso

Future scientific and technological developments in many fields will necessarily depend upon coming to grips with complex systems. Such systems are complex in both their composition–typically many different kinds of components interacting simultaneously and nonlinearly with each other and their environments on multiple levels–and in the rich diversity of behavior of which they are capable.

The Springer Series in Understanding Complex Systems series (UCS) promotes new strategies and paradigms for understanding and realizing applications of complex systems research in a wide variety of fields and endeavors. UCS is explicitly transdisciplinary. It has three main goals: First, to elaborate the concepts, methods and tools of complex systems at all levels of description and in all scientific fields, especially newly emerging areas within the life, social, behavioral, economic, neuro- and cognitive sciences (and derivatives thereof); second, to encourage novel applications of these ideas in various fields of engineering and computation such as robotics, nano-technology, and informatics; third, to provide a single forum within which commonalities and differences in the workings of complex systems may be discerned, hence leading to deeper insight and understanding.

UCS will publish monographs, lecture notes and selected edited contributions aimed at communicating new findings to a large multidisciplinary audience.

More information about this series at http://www.springer.com/series/5394

Rafael Martínez-Guerra
Claudia Alejandra Pérez-Pinacho

Advances in Synchronization of Coupled Fractional Order Systems

Fundamentals and Methods

 Springer

Rafael Martínez-Guerra
Automatic Control
CINVESTAV-IPN
Mexico City, Mexico

Claudia Alejandra Pérez-Pinacho
Automatic Control
CINVESTAV-IPN
Mexico City, Mexico

ISSN 1860-0832   ISSN 1860-0840   (electronic)
Understanding Complex Systems
ISBN 978-3-030-06752-6   ISBN 978-3-319-93946-9   (eBook)
https://doi.org/10.1007/978-3-319-93946-9

© Springer Nature Switzerland AG 2018
Softcover re-print of the Hardcover 1st edition 2018
This work is subject to copyright. All rights are reserved by the Publisher, whether the whole or part
of the material is concerned, specifically the rights of translation, reprinting, reuse of illustrations,
recitation, broadcasting, reproduction on microfilms or in any other physical way, and transmission
or information storage and retrieval, electronic adaptation, computer software, or by similar or dissimilar
methodology now known or hereafter developed.
The use of general descriptive names, registered names, trademarks, service marks, etc. in this
publication does not imply, even in the absence of a specific statement, that such names are exempt from
the relevant protective laws and regulations and therefore free for general use.
The publisher, the authors, and the editors are safe to assume that the advice and information in this
book are believed to be true and accurate at the date of publication. Neither the publisher nor the
authors or the editors give a warranty, express or implied, with respect to the material contained herein or
for any errors or omissions that may have been made. The publisher remains neutral with regard to
jurisdictional claims in published maps and institutional affiliations.

Printed on acid-free paper

This Springer imprint is published by the registered company Springer Nature Switzerland AG
The registered company address is: Gewerbestrasse 11, 6330 Cham, Switzerland

To the memory of my father
who was the source of my teaching
Carlos Martínez Rosales

To my wife and sons
Marilen, Rafael and Juan Carlos.

To my Mother and Brothers
Virginia, Victor, Arturo, Carlos,
Javier and Marisela.

Rafael Martínez-Guerra

To my great family
Miguel, Graciela and Miguel Jr.

To my lovely Husband
Christopher Diego

and specially to
Quico and Max

Claudia Alejandra Pérez-Pinacho

basis for the generalizations encountered in the rest of the book for the same problem. A more detailed description of each chapter is given in the next paragraphs. As a first step towards the comparison of estimation of integer order systems, some preliminary results are stated for commensurate fractional order systems. In Chap. 3, we deal with the synchronization problem of Lorenz system using a proportional reduced-order observer design in the algebraic and differential setting. We prove the asymptotic stability of the resulting error system and by means of algebraic manipulations, we obtain the estimates of the current states (master system). In this chapter, the construction of a proportional reduced-order observer is the main ingredient in our approach. Finally, we present simulations to illustrate the effectiveness of the suggested approach. In Chap. 4, the Immersion and Invariance (I&I) methodology in order to design a novel observer to solve the chaotic synchronization problem for the master–slave configuration is given, where the master belongs to a class of feedback linearized systems. This chapter shows a robust observer which asymptotically calculates the underlying dynamics of the master system and we show convincing numerical simulations that illustrate the effectiveness of the methodology. Chapter 5 deals with the master–slave synchronization scheme for partially known nonlinear fractional order systems, where the unknown dynamics are considered as the master system and we propose the slave system structure which estimates the unknown state variables. Besides, a new fractional model free reduced-order observer inspired on the new concept of Fractional Algebraic Observability (FAO) is introduced; we applied the results to a Rössler hyperchaotic fractional order system and Lorenz fractional order system, and by means of some simulations, we show the effectiveness of the suggested approach. In Chap. 6, we propose the fractional observers for nonlinear commensurate fractional order systems such as a reduced-order observer and a fractional Luenberger Observer in the algebraic and differential setting. We introduce the Fractional Algebraic Observability (FAO) property, like a measurement degree of fractional observability of states variables. Finally, a comparison between two observers illustrates the effectiveness of the suggested techniques, this is performed with two different numerical examples: a linear mechanical oscillator with an integer and a fractional order damping, and a nonlinear fractional order Duffing System. In Chap. 7, a master–slave configuration of strictly different commensurate fractional order Liouvillian systems and the generalized synchronization problem of multiple decoupled families of Liouvillian systems are addressed. The main key ingredient is to find canonical forms for the original systems from a family of fractional differential primitive elements based on the output of each system, taking into account the Liouvillian feature. Fractional order dynamical controllers are designed to solve the generalized multi-synchronization problem. Moreover, it is shown that adding diffusive coupling terms in the dynamical controllers solves the synchronization problem with complex interaction between slave systems, with any type of interplay. Finally, some numerical examples show the effectiveness of the proposed approach. In Chap. 8, we present a new estimator model free type for synchronization of a certain class of incommensurate fractional order systems. We apply our proposals in the master–slave synchronization scheme, where the unknown dynamics are considered as the master system and we propose an observer structure like slave system which estimates the unknown state variables. For solving this problem, we introduce a new Incommensurate Fractional Algebraic Observability (IFAO) property that is used in the design of the slave system. Some numerical results show the effectiveness of the suggested methodology. In Chap. 9, we introduce the Fractional Generalized quasi-Synchronization (FGqS) problem for nonlinear incommensurate chaotic systems in a master–slave configuration, this phenomena is studied from an algebraic and differential point of view that allows us to construct an Incommensurate Fractional Generalized Observability Canonical Form (IFGOCF) from an adequate choice of a differential fractional primitive element. The former enables to design an incommensurate fractional order dynamical controller which is able to achieve synchronization of strictly different incommensurate fractional order chaotic systems. Also, we give the Algebraic Observability Property for incommensurate Fractional Order Systems (IFAO). The process of FGqS is shown with numerical results over Chua-Hartley and Rössler incommensurate fractional order chaotic systems. Finally, in Chap. 10, the problems of synchronization and anti-synchronization are solved for commensurate and incommensurate fractional chaotic systems. A reduced-order fractional integral observer is proposed for fractional systems satisfying the fractional algebraic observability condition. This observer is used as a slave system, whose states are synchronized with the ones from the chaotic system, which acts as a master. The observer uses a reduced set of measurable signals from the master system, solving also the anti-synchronization problem as a straightforward extension of the synchronization one. It is proven that the proposed observer is Mittag-Leffler stable. Numerical simulations on the fractional Lorenz and Rössler systems assess the performance of the proposed methodology.

In this book, several topics of synchronization and secure communication theory are merged and directed towards a constructive solution of the dynamical feedback stabilization problem and some analytic, algebraic, geometric and asymptotic concepts are assembled as design tools for a wide variety of chaotic systems. Differential algebraic concepts show important structural properties of chaotic systems.

It should be noted that the difference with the existing literature of this new book, within the context of coupled fractional order systems, is that this manuscript covers some sensible areas such as stability, estimation, and (anti-)synchronization that have not been given for this type of systems.

In the end, the authors are grateful to referees for a careful and helpful review of the manuscript.

Mexico City, Mexico   Rafael Martínez-Guerra
March 2018   Claudia Alejandra Pérez-Pinacho

Contents

Symbols and Acronyms

$\triangleq$	by definition
$\approx$	approximately equal
$:=$	defined as
$\equiv$	equivalent to
$\neq$	different to
$<\ (>)$	less (greater) than
$\leq\ (\geq)$	less (greater) than or equal to
$\forall$	for all
$\in$	belongs to
$\subset$	subset of
$\cup$	union of sets
$\{\cdot\}$	set
$\sum$	summation
$\mathbb{R}$	the set of real numbers
$\mathbb{R}^+$	positive real numbers
$\mathbb{R}^{m\times n}$	the set of all $m \times n$ matrices with elements from $\mathbb{R}$
$\mathbb{R}^{m\times n}_+$	the set of all $m \times n$ matrices with elements from $\mathbb{R}^+$
$det\ A$	the determinant of a square matrix $A \in \mathbb{R}^{n\times n}$
A^T	the transpose of a matrix obtained by interchanging the rows and columns of A
$(\cdot,\cdot)^T$	the transpose of a vector
$(\cdot,\cdot)'$	the transpose of a vector
max	maximum
min	minimum
sup	supremum, the least upper bound
inf	infimum, the greatest lower bound
B_r	the ball $\{x \in \mathbb{R}^n | \|x\| < r\}$
$f : S_1 \rightarrow S_2$	a function f mapping a set S_1 into a set S_2
$\lambda_{max}(P)(\lambda_{min}(P))$	the maximum (minimum) eigenvalue of a symmetric matrix P
$P > 0$	a positive definite matrix P

$\dot{y}$	the first derivative of y with respect to time
$y^{(i)}$	the ith derivative of y with respect to time
lim	limit
$\|a\|$	absolute value of a scalar a
∞	infinity
$\square$	designation of the end of proofs
$K\langle u \rangle$	differential field generated by the field K, the input $u(t)$ and the time derivative of u
$K\langle u, y \rangle$	differential field generated by the field K, the input $u(t)$, the output $y(t)$, and the time derivatives of u and y
$\hat{x}$	estimated value of x
GS	Generalized Synchronization
AOC	Algebraic Observability Condition
FAO	Fractional Algebraic Observability
PV	Picard–Vessiot
GOCF	Generalized Observability Canonical Form
FGOCF	Fractional Generalized Observability Canonical Form
$D^{(r\alpha)}x(t)$	fractional derivative of Caputo
ODE	Ordinary Differential Equation
K/k	Differential field extension, $k \subset K$
$\bar{B}_r$	Compact Set
$\|\cdot\|_p$	l_p Norm
l_2	Euclidean norm
$\otimes$	Kronecker Product
$\|A\|$	$\sqrt{\lambda max(A*A)}$, $A \in \mathbb{R}^n \times \mathbb{R}^n$
$\|x\|_\infty$	$max\{|x_1|, \ldots, |x_n|\}$
FGqS	Fractional Generalized quasi-Synchronization
$(u, v) = 1$	u, v relatively primes
MFGOCF	Multi-output Fractional Generalized Observability Canonical Form
$\prod$	Product
class C function	These functions are called continuous differentiable

List of Figures

Chapter 1
Introduction

1.1 A Brief History of Fractional Calculus

The origin of the fractional calculus dates back to the late seventeenth century, almost on par with the development of differential and integral calculus, when Isaac Newton and Gottfried Leibniz founded the concept of ordinary derivative. It was in the wake of the introduction of the Leibniz notation

$$\frac{d^n}{dx^n} f(x),$$

which denotes the nth derivative of the function f, when Guillaume de L'Hôpital on September 30, 1695 wrote a letter to Leibniz questioning him about what would happen if n was $1/2$? [1–3]. Thus, becoming the first occurrence of what is now known as fractional calculus; that name comes from the fact that the number $1/2$ is a fraction and which has remained in use since then, although now it is well known that n is not restricted to take values from the set of rational numbers but also from the irrational and complex numbers.

It is known that in 1812, P.S. Laplace wrote some expressions for certain fractional derivatives [4]. And in 1819, Sylvestre F. Lacroix was the first to develop the expression $d^{1/2}x = 2\sqrt{x}/\sqrt{\pi}$ using the fact that $\Gamma(1/2) = \sqrt{\pi}$, where Γ is known as the gamma function whose notation was introduced by AM Legendre [5].

However, probably the first application of the fractional calculus took place in 1823, when the mathematician Niels Henrik Abel used derivatives of arbitrary order in the solution of an integral that emerged from the formulation of the tautochrone [6] (or better known as the problem of the isochronous curve). This problem consists in finding the shape of the curve in which any particle in any position of the curve, when sliding without friction and with uniform gravity, invests the same time in reaching the lowest point of the curve, that is, the time in which it takes a particle to descend to its final position is the same and is independent of its initial position. The integral with which he worked was

© Springer Nature Switzerland AG 2018
R. Martínez-Guerra and C. A. Pérez-Pinacho, *Advances in Synchronization of Coupled Fractional Order Systems*, Understanding Complex Systems, https://doi.org/10.1007/978-3-319-93946-9_1

$$\int_0^x (x-t)^{-1/2} f(t) dt.$$

Now, it is well known that it looks like the fractional order integral used by Riemann.

In 1832 begins the first important study of the fractional calculus by Joseph Liouville to whom today we owe the first definition of fractional derivative [7]. In 1850, with all the advances of the concept of fractional derivative, W. Center analyzed the derivative of the unit and observed differences between the definitions that existed about the derivative of a constant, where the definition of Peacock gave that the result was different to zero, unless the constant was zero, while Liouville's definition gave the derivative zero. Starting from the premise that the two derivative definitions of a constant were true, what does $(d/dx)^\theta x^0$ mean, when θ is a fractional number? [8].

In 1847, Bernhard Riemman developed a theory of fractional operators [9], inspired by a generalization of Taylor's expansion, and he obtained the following definition of fractional integral:

$$\frac{d^{-r}}{dx^{-r}} u(x) = \frac{1}{\Gamma(r)} \int_c^x (x-k)^{r-1} u(k) dk + \phi(x),$$

where he thought it was correct to add a complementary function $\phi(x)$ to the definition, and later Liouville would take Riemann's work to make modifications to the integral (the complementary function taken to be identically zero and the lower limit of integration c is normally zero) that would be known as Riemman–Liouville integral, which would be the first definition of the iterated integral in fractional calculus.

The study of fractional calculus was studied by great mathematical minds such as Euler [10], Lagrange [11], Fourier [12], among the many who studied the fractional calculus and its mathematical consequences [13]. Many found, by using their own notation and methodology, definitions that fit the concept of an integral of non-integer or derived order.

With the advent of the twentieth century and the new mathematical discoveries, many of the scientists who continued to work in the area of fractional calculus found new definitions for the derivative and integral. It was thus that in 1967 M. Caputo gave a new definition of fractional derivative where the derivative of a constant takes the usual interpretation [14].

From the above, the tools developed in the Fractional Calculus had important consequences in the study of Volterra integral equations (related to differential equations of fractional order) with an important potential that led to the first applications of the fractional calculus in control theory. These occurred in the early 60s and made use of the fractional integral operator in the control of servomechanisms and saturation systems. For a more complete treatment of classic historical surveys, the reader is referred to [1, 3].

1.2 Synchronization, Chaos, and Fractional Calculus

1.2.1 Synchronization

Synchronization is ubiquitous in nature, and it is encountered in various fields of science: biology, chemistry, physics, mathematics, astronomy, engineering, social behavior, and technology [15–17]. In a unified attempt to understand this phenomenon, its study relies on dynamical systems of (quasi-)periodic or chaotic behavior[1] in which the influence of one system to the other allows the existence of a common motion, so-called coupled oscillators. More important, applications of the synchronization phenomenon of coupled dynamical systems have a tremendous impact in technology.

The oldest documented study/observation on synchronization of oscillators is a phenomenon that was originally discovered by Christian Huygens in 1665 [23] from his observations of two maritime pendulum clocks hanging on the same wall [17]. This phenomenon occurs when oscillatory (or repetitive) systems via some kind of interaction adjust their behaviors relative to one another so as to attain a state where they work in unison [16]. Roughly speaking, it is said that two dynamical systems are synchronized once the trajectories of one of them are following the another due to an external forcing or coupling between them. In the case of the pendulum clocks, they were the dynamical systems coupled by the wall where they hang on. The result was described by Huygens as some "odd kind of sympathy" [23] now so-called synchronization. As this idea applies to more complex synchronization scenarios, the former can be considered as the simple case of synchronization. In a more formal expression, it is said that two systems are synchronized if the distance between their states converges to zero as time goes to infinity, regardless the initial conditions of both systems.

1.2.2 Chaos

Chaos theory is a field that studies the behavior of dynamic systems that are highly sensitive to initial conditions, one of them poetically and famously known as the butterfly effect [24]. The name of chaos and the adjective of chaotic are used to describe the temporal behavior of a system of aperiodic behavior, apparently random or noisy. The key word is apparently; under this apparent chaotic randomness is a certain order, given by the equations that describe the system. According to the Poincaré–Bendixson theorem, it is known that chaos cannot occur in continuous nonlinear systems of order less than three [25]; this statement is based on the usual

---

[1]There is clear evidence that consensus problems, for systems ranging from simple node integrators to more complex dynamical systems (e.g., chaotic, hyperchaotic, etc.), are completely related to the synchronization phenomenon [18–20]. Even observers (i.e., state estimators [21]) have a clear connection to synchronization problems in some sense [22].

concept of order, such as the number of states in a system or system integrators. While describing the oscillation of chaotic systems, it is found that it does not correspond to such simple geometrical objects like a limit cycle anymore, but rather to complex structures that are called strange attractors (in contrast to limit cycles that are simple attractors) [17]. Furthermore, chaotic systems have been of great interest in recent years given to their presence in physical systems of interest, such as mechanical, electrical, and meteorological systems, among others [26].

Surprisingly enough and despite this apparent randomness, the seminal paper [27] introduced the idea of synchronized chaos related with two identical systems, which can be coupled in such a way that the solution of one always converges to the solution of the other and still chaotic, independently of initial conditions and parameters. One of the current challenges, since its introduction by Pecora and Carrol [27], is still to explain the synchronization between chaotic systems. The synchronization of chaos appears in many natural processes such as the relationship between neurons and heartbeats, among others. It has been extensively studied, given their applications to areas such as medicine [16], biological systems [28, 29], and whose study of greater relevance consists of understanding the synchronization between different chaotic systems with potential applications in secure communications [19, 30, 31]. This has been lead to several engineering applications; the most important is related to the transmission of information using chaotic systems [21, 32, 33].

For this type of systems, there exist several types of synchronization schemes: Complete Synchronization (CS), Generalized Synchronization (GS), Phase Synchronization, Projective Synchronization, Lag Synchronization, and Observer-based Synchronization. For most of these approaches, it is required that the system meet certain characteristics related to the Lyapunov exponents; that is why it can be applied only to certain kind of chaotic oscillators [16, 17, 27]. A complete review of this topic is beyond the scope of this work, but the reader is referred to: [16, 17, 27, 34–36].

The purpose of this book is to tackle mainly two fundamental problems in chaos synchronization: Generalized synchronization and observer-based synchronization, both are commonly encountered in a unidirectional configuration unless otherwise stated. One of the methods used for solving the synchronization problem in chaotic systems consists of decomposing the system in question in two subsystems, where one of them acts as a driver system in order to synchronize the behavior of the other one. This configuration so-called master–slave synchronization means that, although one system responds to the other, the reciprocal does not happen.

1.2.3 Synchronization and Observation

Chaotic synchronization has been a topic of huge interest due to its engineering applications. Several related works can be found in the literature (see [27, 36–39] and references therein). The research focus into two branches, one related to the application of state observers for synchronizing nonlinear oscillators, and the other related to application of control laws to achieve synchronization between systems with different structures. The first one has aroused great interest [22, 40, 41]. The key idea is to design observers to accomplish chaos synchronization, where the slave is actually an observer coupled to the master through its corresponding output. That is to say, slave estimates the dynamics of the unknown states of master system by means of measurements of its output. Therefore, the chaos synchronization problem can be posed as an observer design in which the coupling signal is seen as the output and the slave system as the observer. Several works related to this have been published.

Different kinds of observers have been used as slave systems, such as the Luenberger [42, 43], high-gain [44], adaptive [45], reduced-order [46], polynomial [47], and bounded error [21]. Recently, the differential algebraic approach has been used for the design of reduced-order observers [48]. Given that these observers do not require to have full knowledge of the dynamics of the system, it is required that the variables to estimate satisfy properties of algebraic observability.

On the other hand, the use of control laws makes it possible to achieve synchronization between nonlinear (chaotic) systems with different structures and orders. In [37, 47], feedback strategies' control and nonlinear observers were proposed. Theoreticians have paid attention to this issue, coming up with interesting works like [49–51]. Other techniques that have attained noteworthy development are active control [52, 53], adaptive control [37, 47, 54, 55], backstepping design [56–58], and sliding mode control [59, 60], among others. There are many mechanisms to understand the synchronization of chaos as stated in the last section. Of greater interest because it allows describing this phenomenon between chaotic systems of different nature is known as Generalized Synchronization (GS). This concept, introduced in [61], is used to describe the synchronization scheme of chaotic systems coupled unidirectionally (i.e., in a master–slave configuration) and occurs when the trajectories of a system (slave system), through a mapping, they are equal to the trajectories of another (master system). When this mapping is equal to identify a particular case known as CS occurs, this is the case where systems possess identical dynamic model. GS problem is twofold: first, find a mapping that relates the trajectories of the slave system to the master system; and second, give an explicit form of the mapping.

1.2.4 Synchronization of Fractional Order Chaotic Systems

Currently, the fractional calculus is present in almost all areas of engineering and science, such as physics, electrical engineering, robotics, control systems, chemistry, and bioengineering, among others, and one of the important areas of application for our case study in the theory of chaos.

Inspired by the above observations and due to the discovery of chaotic behaviors in fractional differential equations [62, 63], where it is now known that the model of a chaotic system can be reordered to three individual differential equations containing derivatives of the non-integer-order (Fractional), extensions to the fractional case on the synchronization of these systems have been an area of great interest among the scientific community. The attractiveness of these systems is evident, and it lies in the order of the derivative as a new design parameter.

Recently, synchronization of fractional order chaotic systems has received much attention as the integer-order case. The former is a broad topic tackled from different techniques: a one-way coupling and a projective synchronization scheme for the unified fractional order system is addressed in [64, 65], respectively. A dynamical analysis for the one-way coupling scheme of fractional order Liu systems is obtained in [66]. Synchronization of hyperchaotic Lorenz system is tackled from an active control technique in [67]. In [68], a linear active control technique is used for synchronization in driver and response configuration, where the proposed methodology is applied to synchronize identical systems with commensurate and incommensurate fractional order. An active sliding mode control is given in [69, 70], and a modified version of [69] is given in terms of the projective synchronization problem in [71]. An adaptive projective synchronization method with parametric uncertainty of fractional Lorenz systems with reduced number of active control signals is given in [72], and it is proven that synchronization errors can only be bounded. In [73], a synchronization method with optimal active control and fractional cost function is proposed. Particularly, in [74, 75], both consider the generalized synchronization problem for fractional order chaotic systems. These two last works are extensions from the auxiliary system approach of Abarbanel et. al. [76], and the differential algebraic approach of Martínez–Guerra et al. [47], respectively. There are clear extensions to the fractional case of this methodology called Fractional Generalized Synchronization (FGS) for a class of systems [47].

The existence of observers for fractional systems has shown another interesting approach, widely used in full-order systems [22], for the solution of the problem of synchronization of fractional chaotic systems. This is to translate the problem of synchronization into an observation problem, and of course, this happens when the trajectories of the master system can be estimated from an observer. In this case, the observer plays the role of the slave system and the error of estimation of the synchronization error.

References

1. Bertram Ross, "The development of fractional calculus 1695-1900", Historia Mathematica vol. 4, pp. 75–89, 1977.
2. Letter from Hanover, Germany to G.F.A. L'Hospital, September 30, 1695 Mathematische Schriften 1849, 2, Olms Verlag, Hildesheim, Germany (1695), pp. 301–302 reprinted 1962.
3. Keith B. Oldhamm, and Jerome Spainer, *"The Fractional Calculus: Theory and applications of Differentiation and Integration to Arbitary Order"*, Dover, 1974.
4. Pierre Simon de Laplace, *"Théorie analytique des probabilités"*, Vol. 7. Courcier, 1820.
5. Silvestre Francois Lacroix, *"Traité du calcul differentiel et du calcul intégral"*, Vol. 1. JBM Duprat, 1797.
6. N.H Abel, *"Solutions de quelques problèmes à l'aide d'intégrales définies"* . Oeuvres complètes, Vol. 1: pp. 11–18, 1823.

7. J. Liouville, *"Mémoire sur le Calcul des différentielles à indices quelconques"*, J Ecole Polytech, Vol 13, Section 21 pp. 71–162, 1832.

8. William Center, *"On the Value of $(d/dx)^\theta x^0$ When θ is a Positive Proper Fraction"*, Cambridge & Dublin Math. Journal 3, pp. 163–169, 1848.

9. Bernhard Riemman, *"Versuch einer allgemeinen Auffassung der Integration und Differentiation"*, Gesammelte Werke 62, 1876.

10. L. Euler, *"De progressionibus Transcentibus, sev Quarum Termini Algebraice Dari Nequeunt"*, Comment. Acad. Sci. Imperialis Petropolitance Vol 5, pp. 38–57, 1738.

11. Joseph Louis Lagrange, *"Sur une nouvelle espèce de calcul rélatif à la différentiation et à l'intégration des quantités variables"*, Académie royale des sciences et belles lettres, 1772.

12. Fourier, Joseph, *"Theorié analytique de la chaleur, par M. Fourier"*, Chez Firmin Didot, père et fils, 1822.

13. Nishimoto Katsuyuki, *"An Essence of Nishimoto's Fractional Calculus (Calculus in the 21st Century): Integrations and Differentiations of Arbitrary Order"*, Descartes Press Company, 1991.

14. Bonilla, B., Kilbas, A. A., and Trujillo, J. J., *"Cálculo Fraccionario y Ecuaciones Diferenciales Fraccionarias"*, Uned, Madrid, 2003.

15. A. Balanov, N. Janson, D. Postnov, O. Sosnovtseva, Synchronization: From Simple to Complex, Springer, 2009.

16. E. Mosekilde, Y. Maistrenko, D. Postnov, Chaotic synchronization application to living systems, World Scientific Publishing, 2002.

17. A. Pikovsky, M. Rosenblum, J. Kurths, Synchronization: A universal concept in nonlinear sciences, Cambridge University Press, 2001.

18. Ljupco Kocarev (Ed.), Consensus and synchronization in complex networks, Springer, 2013.

19. R Martínez-Guerra, C. D. Cruz-Ancona, C. A. Pérez-Pinacho, Generalized multi-synchronization viewed as a multi-agent leader-following consensus problem, Applied Mathematics and Computation 282 (2016) 226–236.

20. C. D. Cruz-Ancona, R. Martínez-Guerra, C. A. Pérez-Pinacho, Generalized multi-synchronization: A leader-following consensus problem of multi-agent system, Neurocomputing, Vol. 233, pp. 52–60, 2017.

21. Rafael Martínez-Guerra, and Christopher Diego Cruz-Ancona. Algorithms of Estimation for Nonlinear Systems, Springer International Publishing AG, 2017.

22. Nijmeijer, Henk, and Iven MY Mareels, *"An observer looks at synchronization"*, IEEE Transactions on Circuits and Systems I: Fundamental theory and applications vol. 44. no. 10, pp. 882–890,(1997)

23. Christiaan Huygens, La lettre á R.F. de Sluse, 24 février 1665. In: Oeuvres complètes. Tome V. Correspondance 1664–1665 (ed. D. Bierens de Haan). Martinus Nijhoff, Den Haag, pp 241, 1893.

24. Edward N. Lorenz, Section of planetary sciences: The predictability of hydrodynamic flow. Transactions of the New York Academy of Science, 25(4 Series II), 409–432, 1963.

25. Strogatz, Steven H. *"Nonlinear dynamics and chaos: with applications to physics, biology, chemistry, and engineering"*. Westview press, 2014.

26. J. Guckenheimer, P. Holmes, Nonlinear Oscillations, Dynamical Systems, and Bifurcations of Vector Fields, Springer, 1993.

27. L. M. Pecora, T. L. Carroll, Synchronization in Chaotic Systems, Physical Review Letters 64(8) (1990) 821–824.

28. V. S. Anishchenko, V. Astakhov, A. Neiman, T. Vadivasova, L. Schimansky-Geier, Nonlinear Dynamics of Chaotic and Stochastic Systems: Tutorial and Modern Developments, Springer, 2007.

29. R. Engbert, F. R. Drepper, Chance and chaos in population biology models of recurrent epidemics and food chain dynamics, Chaos, Solitons & Fractals 4 (1994) 1147–1169.

30. G. Kaddoum, M. Coulon, D. Roviras, P. Chargé, Theoretical performance for asynchronous multi-user chaos-based communication systems on fading channels, Signal Processing 90(11) (2010) 2923–2933.

31. R. Martínez-Guerra, J. J. Montesinos García, S. M. Delfín-Prieto, Secure Communications via Synchronization of Liouvillian Chaotic Systems, Journal of the Franklin Institute 353 (2016) 4384–4399.
32. S. Banerjee (ed.), Applications of Nonlinear Dynamics and Chaos in Science and Engineering, Vol. II (Springer, 2012)
33. L.E. Keshet, Mathematical Models in Biology (Random House, New York, 1988)
34. César Cruz and Henk Nijmeijer. Synchronization through filtering. International Journal of Bifurcation and Chaos, 10(04): 763–775, 2000.
35. Rafael Martínez-Guerra, and Wen Yu, Chaotic synchronization and secure communication via sliding-mode observer, International Journal of Bifurcation and Chaos 18(01): 235–243 (2008).
36. H. Sira-Ramírez and César Cruz-Hernández. Synchronization of chaotic systems: a generalized hamiltonian systems approach. International Journal of Bifurcation and Chaos, 11(05):1381–1395, 2001.
37. Alexander L Fradkov and A Yu Markov. Adaptive synchronization of chaotic systems based on speed gradient method and passification Circuits and Systems I: Fundamental Theory and Applications, IEEE Transactions on, 44(10):905–912, 1997.
38. Ljupco Kocarev, Alain Shang, and Leon O Chua. Transitions in dynamical regimes by driving: a unified method of control and synchronization of chaos. International Journal of Bifurcation and Chaos, 3(02): 479–483, 1993.
39. Chai Wah We and Leon o Chua. A simple way to synchronize chaotic systems with application to secure communication systems. International Journal of Bifurcation and Chaos, 3(06):1619–1627
40. Zhang Hua-Guand, Ma Tie-Dong, Yu Wen, and Fu Jie. A practical approach to robust impulsive lag synchronization between different chaotic systems. Chinese Physics B, 17(10):3616, 2008.
41. Zhang Hua-Guang, Zhao Yan, Yu Wen, and Yang Dong-Sheng. A unified approach to fuzzy modelling and robust synchronization of different hyperchaotic systems. Chinese Physics B, 17(11):4056, 2008.
42. M. Baumann, R. I. Leine, Synchronization-based state observer for impacting multibody systems using switched geometric unilateral constraints, Proceedings of the ECCOMAS Thematic Conference on Multibody Dynamics, June 29–July 2 2015, Barcelona, Catalonia, Spain, p. 585–596.
43. S. Poinsard, A. Lor, Robust Communication-Masking via a Synchronized Chaotic Lorenz Transmission System, in J.A. Tenreiro-Machado, Albert C.J. Luo, R. S. Barbosa, M. F. Silva, L. B. Figueiredo (Eds.), Nonlinear Science and Complexity, Springer, 2011, p. 357–365.
44. P. Canyelles-Pericas, K. Busawon, High gain observer with algorithm transformation to extended Jordan observable form for chaos synchronization applications, Proceedings of the 2014 UKACC International Conference on Control, 9th–11th July 2014, Loughborough, U.K., p. 262–267.
45. H. Dimassi, A. Loria, S. Belghith, Adaptive observers-based synchronization of a class of Lur'e systems with delayed outputs for chaotic communications, Proceedings of the 2012 IFAC Conference on Analysis and Control of Chaotic Systems, June 20–22 2012, Cancún, México, p. 255–260.
46. E. Solak, A reduced-order observer for the synchronization of Lorenz systems, Physics Letters A 325 (2004) 276–278.
47. Rafael Martínez-Guerra, G. C. Gómez-Cortés, and C. A. Pérez-Pinacho. Synchronization of integral and fractional order chaotic systems. Springer, 2015.
48. R. Martínez-Guerra, A. Luviano-Juárez, J. J. Rincón-Pasaye, On nonlinear observers: A differential algebraic approach, Proceedings of the 2007 American Control Conference, July 11–13 2007, New York City, USA, p. 1682–1686.
49. Alexander Fradkov, Henk Nijmeijer and Alexey Markov. Adaptive Observer.based synchronization for communication. International Jorunal of Bifurcation and Chaos, 10(12):2807–2813, 2000.
50. A Yu Pogromsky. Passivity based design of synchronizing systems. International Journal of Bifurcation and Chaos, 8(02):295–319, 1998.

51. Alexander Pogromsky, Giovanni Santoboni, and Henk Nijmeijer. Partial synchronization: from symmetry towards stability. Physica D: Nonlinear Phenomena, 172(1): 65–87, 2002.
52. Ahmet Ucar, Karl E Lonngren, and Er-Wei Bai. synchronization of the unified chaotic systems via active control. Chaos, solitons & Fractals, 27(5):1292–1297, 2006.
53. UE Vicent. Synchronization of rikitake chaotic attractor using active control. Physics Letters A, 343(1):133–138, 2005
54. Shihua Chen and Jinhu Lü. Synchronization of an uncertain unified chaotic system via adaptive control. Chaos, Solitons & Fractals, 14(4):643–647, 2002.
55. Moez Feki. An adaptive chaos synchronization scheme applied to secure communication. Chaos, Solitons & Fractals, 18(1): 141–148, 2003.
56. Chih-Min Lin, Ya-Fu Peng, and Ming-Hung Lin, Cmac-based adaptive backstepping synchronization of uncertain chaotic systems. Chaos, solitions & and fractals, 42(2):981–988, 2009.
57. Yongguang Yu and Suochun Zhang. Adaptive backstepping synchronization of uncertain chaotic system. Chaos, Solitons & Fractals, 21(3):643–649, 2004.
58. Jian Zhang, Chunguang Li, Hongbin Zhang and Juebang Yu. Chaos Syncrhonization using single variable feedback based on backstepping method. chaos, Solitons & Fractals, 21(5):1183–1193, 2004.
59. Her-Terng Yau. Design of adaptive sliding mode controller for chaos synchronization with uncertainties. Chaos, Solitons & Fractals, 22(2):347, 2004.
60. Hao Zhang, Xi-Kui Ma, and Wei-Zeng Liu. Synchronization of Chaotic systems with parametric uncertainty using active sliding mode control Chaos, Solitons & Fractals, 21(5): 1249–1257, 2004.
61. B.R. Hunt, E. Ott, J.A. Yorke, *"Diferentiable generalized synchronization of chaos"*, Phys Rev E; 55:4029, 1997.
62. Mohammad Saleh Tavazoei, Mohammad Haeri, *"A necessary condition for double scroll attractor existence in fractional-order systems"*, Physics Letters A 367, pp. 102–113, 2007.
63. Mohammad Saleh Tavazoei, Mohammad Haeri, *"Chaotic attractors in incommensurate fractional order systems"*, Physica D 237 , pp. 2628–2637, 2008.
64. Wu, X., Li, J., Chen, G., Chaos in the fractional order unified system and its synchronization. Journal of the Franklin Institute, 2008; 345(4): 392–401.
65. Wang, X.Y., & He, Y. Projective synchronization of fractional order chaotic system based on linear separation. Physics Letters A, 2008; 372(4): 435–441.
66. Wang, X. Y., & Wang, M. J. Dynamic analysis of the fractional-order Liu system and its synchronization. Chaos: An Interdisciplinary Journal of Nonlinear Science, 2007; 17(3): 033106.
67. Wang, X. Y., & Song, J. M. Synchronization of the fractional order hyperchaos Lorenz systems with activation feedback control. Communications in Nonlinear Science and Numerical Simulation, 2009; 14(8): 3351–3357.
68. Odibat, Z. M., Corson, N., Aziz-Alaoui, M. A., & Bertelle, C., Synchronization of chaotic fractional-order systems via linear control. International Journal of Bifurcation and Chaos, 2010, 20(1):81–97.
69. M.S. Tavazoei, M. Haeri, Synchronization of chaotic fractional-order systems via active sliding mode controller, Physica A: Statistical Mechanics and its Applications. 2008; 387(1): 57–70.
70. A. Razminia, D. Baleanu, Complete synchronization of commensurate fractional order chaotic systems using sliding mode control, Mechatronics. 2013; 23(7): 873–879.
71. Wang, X. Y., Zhang, X., & Ma, C. Modified projective synchronization of fractional-order chaotic systems via active sliding mode control. Nonlinear Dynamics, 2012; 69(1): 511–517.
72. N. Aguila-Camacho, M. A. Duarte-Mermoud, & E. Delgado-Aguilera, Adaptive synchronization of fractional Lorenz systems using a reduced number of control signals and parameters. Chaos, Solitons & Fractals, 2016; 87: 1–11.

73. Behinfaraz, R., & Badamchizadeh, M. Optimal synchronization of two different in commensurate fractional-order chaotic systems with fractional cost function. Complexity, 2016, 21(S1): 401–416.
74. Deng, W., Generalized synchronization in fractional order systems, Physical review E. 2007; 75(5): 056201.
75. R. Martínez-Guerra, & J. L. Mata-Machuca, Fractional generalized synchronization in a class of nonlinear fractional order systems, Nonlinear Dynamics. 2014; 77(4): 1237–1244.
76. Henry D. I. Abarbanel, Nikolai F. Rulkov, Mikhail M Sushchik, Generalized synchronization of chaos: The auxiliary system approach, Phys. Rev. E. , 52(1): 214–217, 1995.

Chapter 2
Basic Concepts and Preliminaries

In this chapter, definitions and concepts about the fractional calculus are presented. Some of the concepts and definitions have been divided into two parts in order to better locate the topic of interest, such as the concepts of commensurate and incommensurate systems. Taking into account that there are tools that serve the same for both cases.

Let us start by defining the n-th derivative of a function x for fractional case.

A fractional derivative of order $\alpha \in \mathbb{R}$ is an operator that generalizes the ordinary derivative such that

$$x^{(\alpha)} = D^{\alpha} f(x) = \frac{d^{\alpha} f(x)}{dx^{\alpha}},$$

when $\alpha = 1$ matches the ordinary differential operator.

Now, remember the factorial of a number n

$$n! = \prod_{k=1}^{n} k,$$

where $n \in \mathbb{N} > 0$, so what happens with the rational, irrational, and complex numbers? Let us see the following definition.

2.1 Gamma Function

In the fractional calculus, one of the functions that will appear most in the tools of the fractional calculus is the Gamma Function [1].

© Springer Nature Switzerland AG 2018

R. Martínez-Guerra and C. A. Pérez-Pinacho, *Advances in Synchronization of Coupled Fractional Order Systems*, Understanding Complex Systems, https://doi.org/10.1007/978-3-319-93946-9_2

This function generalizes the factorial expression $n!$.

$$\Gamma(z) = \int_0^\infty e^{-t} t^{z-1} dt \tag{2.1}$$

Considering z to be a real number, the above statement implies that Gamma function is defined continuously for positive real values of Z.

Example 2.1 • $\Gamma(1) = 1$

$$\Gamma(1) = \int_0^\infty e^{-t} dt = [-e^{-t}]_0^\infty = 1$$

• $\Gamma(\frac{1}{2}) = \sqrt{\pi}$

$$\Gamma(\frac{1}{2}) = \int_0^\infty t^{1/2} e^{-t} dt = 2\frac{\sqrt{\pi}}{2} = \pi$$

With a change of variable $u = t^{1/2}$ with $du = \frac{1}{2} t^{-1/2} dt$ and along with the known function of the Gaussian Integral $\int_{-\infty}^\infty e^{-x^2} dx = \sqrt{\pi}$.

Now, we can introduce the following special function.

2.2 Mittagg–Leffler Function

The Mittag–Leffler (ML) function is the basis function of fractional calculus, as the exponential function is to the integer order calculus [2].
  One-Parameter Mittag–Leffler Function

$$E_\alpha(z) = \sum_{k=0}^\infty \frac{z^k}{\Gamma(\alpha k + 1)}, \tag{2.2}$$

where Γ is the already seen Gamma function
  Two-Parameter Mittag–Leffler Function [3]

$$E_{\alpha,\beta}(z) = \sum_{k=0}^\infty \frac{z^k}{\Gamma(\alpha k + \beta)} \quad (\alpha > 0, \beta > 0) \tag{2.3}$$

Note that if $\beta = 1$, two-parameter ML function becomes one-parameter ML function. This function is used to solve fractional differential equations as the exponential function in integer order systems. In the particular case when $\alpha = \beta = 1$, we have that $E_{1,1}(z) = e^z$. Now if we have particular values of α, the function (2.3) has asymptotic behavior at infinity.

Definition 2.1 ([4]) The solution of system $D^\alpha x(t) = f(t, x)$ is said to be Mittag–Leffler stable if

$$\|x(t)\| \leq \{m[x(0)]E_{\alpha,1}(-\lambda t^\alpha)\}^b,$$

$\alpha \in (0, 1)$, $\lambda \geq 0$, $b > 0$, $m(0) = 0$, $m(x) \geq 0$, and $m(x)$ is locally Lipschitz (with Lipschitz constant m_0) on $x \in \mathbb{B}$, an open subset of $\mathbb{R}^n$.

Mittag–Leffler function satisfy the following Theorem:

Theorem 2.1 ([5]) *If $\alpha \in (0, 2)$, β is an arbitrary complex number and μ is an arbitrary real number such that*

$$\frac{\pi \alpha}{2} < \mu < \min\{\pi, \pi\alpha\}, \tag{2.4}$$

then, for an arbitrary integer $\kappa \geq 1$ the following expansion holds:

$$E_{\alpha,\beta}(z) = -\Sigma_{i=1}^{\kappa} \frac{1}{\Gamma(\beta - \alpha i)z^i} + O\left(\frac{1}{|z|^{\kappa+1}}\right), \tag{2.5}$$

with $|z| \to \infty$, $\mu \leq |\arg(z)| \leq \pi$. $\qquad\qquad\square$

The Mittag–Leffler function has the following properties:

Property 2.1 ([5])

$$\int_0^t \tau^{\beta-1} E_{\alpha,\beta}(-k\tau^\alpha)d\tau = t^\beta E_{\alpha,\beta+1}(-kt^\alpha), \beta > 0.$$

Property 2.2 ([5])
$E_{\alpha,\beta}(-x)$ is completely monotonic, i.e., $(-1)^n E_{\alpha,\beta}^{(n)}(-x) \geq 0$ for $0 < \alpha \leq 2$ and $\beta \geq \alpha$, for all $x \in (0, \infty)$ and $n \in \mathbb{N} \cup \{0\}$.

2.3 Fractional Operators

There are several definitions of a non-integer derivative of order α, commonly known as fractional order derivative operators (see [5–7]). In this book, the Riemman–Liouville fractional operator and the Caputo fractional operator are used.

2.3.1 Riemman–Liouville Fractional Operator

Definition 2.2 (*Riemman–Liouville Fractional Integral*) Let $\alpha \in \mathbb{R}^+$ and let f be piecewise continuous on $J' = (0, \infty)$ and integrable on any finite subinterval of $J = [0, \infty)$ (functions of class C). Then, for $t > 0$ we call

$$x^{(-\alpha)} = {}_0D_t^{-\alpha}x(t) = \frac{1}{\Gamma(\alpha)} \int_0^t (t - \tau)^{\alpha-1} x(\tau)d\tau, \qquad (2.6)$$

the Riemman–Liouville fractional integral of x of order α.   □

There exist several definitions of fractional derivatives of α order [6–9]. However, here we will use the Riemman–Liouville approach.

Definition 2.3 (*Riemman–Liouville Fractional Derivative* [5]). Let f be a function of class C and let $\mu > 0$. Let m be the smallest integer that is greater or equal to μ. Then, the fractional derivative of f of order μ (if it exists) is defined as

$$D^\mu f(t) = D^m\{D^{-\nu}f(t)\} \;, \mu > 0, \; t > 0, \qquad (2.7)$$

where $\nu = m - \mu \geq 0$ □

Note that $\nu = 0$ implies $D^\mu f(t) = D^m f(t)$
Now, we define a sequential operator, as follows.

$$D^{r\alpha}x(t) = \underbrace{[D_t^\alpha[D_t^\alpha...[D_t^\alpha x(t)]]]}_{r-times}, \qquad (2.8)$$

i.e., the Riemman–Liouville fractional derivative of order α applied r-times sequentially $r \in \mathbb{N}$, with $D^0 x(t) = x(t)$, we can note that if $r = 1$, then $D^\alpha x(t) = x^{(\alpha)}$

2.3.2 Caputo Fractional Operator

The Caputo fractional derivative of order $\alpha \in \mathbb{R}^+$ of a function x is defined as (see [10])

$$x^{(\alpha)} = {}_{t_0}D_t^\alpha x(t) = \frac{1}{\Gamma(m - \alpha)} \int_{t_0}^t \frac{d^m x(\tau)}{d\tau^m}(t - \tau)^{m-\alpha-1}\, d\tau, \qquad (2.9)$$

where $m - 1 \leq \alpha < m$, $\frac{d^m x(\tau)}{d\tau^m}$ is the m-th derivative of x, $m \in \mathbb{N}$ and Γ is the gamma function[1].

---

[1]To simplify the notation, we omitted the time dependence on $x^{(\alpha)}$, in what follows we take $t_0 = 0$.

Now, we define a sequential operator (see [7]) as follows

$$\mathscr{D}^{(r\alpha)}x(t) = \underbrace{{}_{t_0}D_t^\alpha \, {}_{t_0}D_t^\alpha \cdots {}_{t_0}D_t^\alpha \, {}_{t_0}D_t^\alpha}_{r\text{-times}} \, x(t), \tag{2.10}$$

with $r \in \mathbb{N}$, note that $\mathscr{D}^{(0)}x(t) = x(t)$, $\mathscr{D}^{(\alpha)}x(t) = x^{(\alpha)}$ for $r = 0$ and $r = 1$, respectively.

2.3.2.1 General Example

Example 2.2 Now that we know the Caputo operator, the gamma function and the ML function, we can obtain the solution of the following fractional system.

$$x^{(\alpha)}(t) = ax(t), \quad x(0) = x_0,$$

assuming the type solution

$$x(t) = a_0 + a_1 t^\alpha + a_2 t^{2\alpha} + \dots + a_k t^{k\alpha}, \tag{2.11}$$

We will find the coefficients of the expansion in power series.

At the initial instant $t = 0$, the expansion in power series has as a result

$$x(0) = a_0,$$

Applying the derivative of Caputo to the expansion in series of powers and using property

$$D^\alpha t^\sigma = \frac{\Gamma(\sigma + 1)}{\Gamma(\sigma + 1 - \alpha)} t^{\sigma - \alpha},$$

we obtain

$$x^{(\alpha)}(t) = 0 + a_1 \frac{\Gamma(\alpha + 1)}{\Gamma(1)} + a_2 \frac{\Gamma(2\alpha + 1)}{\Gamma(\alpha + 1)} t^\alpha$$
$$+ \dots + a_k \frac{\Gamma(k\alpha + 1)}{\Gamma(1 + (k-1)\alpha)} t^{\alpha(k-1)} = ax(t),$$

for $t = 0$, we have

$$a_1 = ax(0) \frac{\Gamma(1)}{\Gamma(\alpha + 1)} = a \frac{x(0)}{\Gamma(\alpha + 1)},$$

applying successive derivatives of (2.11) and substituting $t = 0$ and we find the coefficients a_n, $n = 1, \dots, k, \dots$ such that

$$x^{(2\alpha)}(t) = 0 + a_2 \frac{\Gamma(2\alpha + 1)\Gamma(\alpha + 1)}{\Gamma(\alpha + 1)\Gamma(1)}$$

$$+ \dots + a_k \frac{\Gamma(k\alpha + 1)\Gamma(1 + \alpha(k - 1))}{\Gamma(1 + (k - 1)\alpha)\Gamma(1 + \alpha(k - 2))} t^\alpha(k - 2) + \dots$$

$$= ax^{(\alpha)}(t) = (a)ax(t) = a^2 x(t),$$

if $t = 0$

$$a_2 = \frac{a_2 x(0)}{\Gamma(2\alpha + 1)}$$

For the k-th term

$$x^{(k\alpha)}(t) = a_k \Gamma(k\alpha + 1)t^0 = a_k \Gamma(k\alpha + 1) = ax^{((k-1)\alpha)}(t)$$

$$= (a)a^{k-1}x(t) = a^k x(t),$$

if $t = 0$

$$a_k = \frac{a^k x(0)}{\Gamma(k\alpha + 1)}$$

Finally, replacing all the terms in the series

$$x(t) = x(0) + a \frac{x(0)}{\Gamma(\alpha + 1)} t^\alpha + a^2 \frac{x(0)}{\Gamma(2\alpha + 1}t^{2\alpha} + \dots$$

$$+ \frac{a^k x(0)}{\Gamma(k\alpha + 1)}t^{k\alpha} + \dots = \Sigma_{k=0}^\infty \frac{a^k t^{k\alpha}}{\Gamma(k + \alpha + 1)}x(0),$$

Note that the previous expression corresponds to a function of Mittag–Leffler

$$x(t) = E_{\alpha,1}(at^\alpha)x(0),$$

which is solution of $x^{(\alpha)}(t) = ax(t)$.

2.4 Laplace Transform of Fractional Integrals and Fractional Derivatives

The Laplace transform is a powerful method in the study of fractional differential–integral equations. In the following paragraphs are introduced some facts about the use of the Laplace transform in fractional calculus [7].

Let f a function of class C, if f is of exponential order, its fractional integral Laplace transform is given by

$$\mathscr{L}\left\{D^{-\alpha}f(t)\right\} = \frac{1}{\Gamma(\alpha)}\mathscr{L}\{f(t)\} = s^{-\alpha}F(s) , \quad \alpha > 0. \qquad (2.12)$$

Let us assume that the Laplace transform of $f(t)$ exists and is denoted by $F(s)$, Thus, the following equation holds

$$\mathscr{L}\{D^{\alpha}f(t)\} = s^{\alpha}F(s) - \sum_{k=0}^{m-1} s^{m-k-1} D^{k-m+\alpha} f(0), \tag{2.13}$$

where $m - 1 < \alpha \leq m$, for $m \in \mathbb{N}$.

2.5 Existence and Uniqueness of Fractional Order Systems

Next, we present some sufficient conditions, in the scalar case, to the existence and uniqueness of solutions of fractional differential equations with Caputo derivative (see [9] for more details) as a generalization of the integer order counterpart. Extensions to vector cases are easily extended from this.

Consider the nonlinear differential equation of fractional order $\alpha > 0$ and $a \leq x \leq b$

$$(^{C}D_{a+}^{\alpha}y)(x) = f[x, y(x)] , \tag{2.14}$$

where $(^{C}D_{a+}^{\alpha}y)(x)$ as a Caputo fractional derivative on a finite interval $[a, b]$, with the initial conditions

$$y^{(k)}(a) = b_k, \quad b_k \in \mathbb{R}, \tag{2.15}$$

with $k = 0, 1, ..., n - 1$; $n = -[-\alpha]$. Considering a fractional differential equation with $0 < \alpha < 1$, and $f(x, y)$ is bounded by $\bar{G}$ in $\mathbb{R} \times \mathbb{R}$ and satisfies the Lipschitz condition with respect to y.

$$|f[x, y_1] - f[x, y_2]| \leq A|y_1 - y_2|, \tag{2.16}$$

where A is a constant $A > 0$ does not depend on x.

The solution of (2.14) over the interval defined by $a \leq x \leq b$ is understood as a continuously differentiable function within some defined space associated with its Volterra integral equation

$$y(x) = \sum_{j=0}^{n-1} \frac{b_j}{j!}(x - a)^j + \frac{1}{\Gamma(\alpha)} \int_a^x \frac{f[t, y(t)]}{(x - t)^{1-\alpha}} dt \quad (a \leq x \leq b), \tag{2.17}$$

such that $(^{C}D_{a+}^{\alpha}y)(x)$ is well defined and (2.14) fulfills for all $a \leq x \leq b$ with initial conditions at the end point of the interval. The latter is a generalization of the Cauchy problem for ordinary differential equations (ODE) with $\alpha = n$, $n \in \mathbb{N}$ for a suitable function $y(x)$:

$$y^{(n)}(x) = f[x, y(x)], \quad (a \le x \le b), \quad y^{(k)}(a) = b_k \in \mathbb{R} \quad (k = 0, 1, ..., n-1).$$
$$(2.18)$$

Before stating the existence and uniqueness theorem, consider the next spaces of functions. Let $\Omega = [a, b](-\infty \le a < b \le \infty)$ and $m \in \mathbb{N}_0 = \{0, 1, ...\}$. We denote by $C^m(\Omega)$ a space of functions f which are m times continuously differentiable on Ω with the norm

$$||f||_{C^m} = \Sigma_{k=0}^m ||f^{(k)}||_C = \Sigma_{k=0}^m \max_{x \in \Omega} |f^{(k)}(x)|, \quad m \in \mathbb{N}_0.$$

In particular, for $m = 0$, $C^0(\Omega) \equiv C(\Omega)$ is the space of continuous functions f on Ω with the norm

$$||f||_C = \max_{x \in \Omega} |f(x)|,$$

when $\Omega = [a, b]$ is a finite interval and $\gamma \in \mathbb{C}(0 \le \Re(\gamma) < 1)$, we introduce the weighted space $C_\gamma[a, b]$ of functions f given on $(a, b]$, such that the function $(x - a)^\gamma f(x) \in C[a, b]$, and

$$||f||_{C_\gamma} = ||(x-a)^\gamma f(x)||_C, \quad C_0[a, b] = C[a, b].$$

For $n \in \mathbb{N}$, we denote by $C_\gamma^n[a, b]$, the Banach space of function $f(x)$ which are continuously differentiable on $[a, b]$ up to order $n - 1$ and have the derivative $f^{(n)}(x)$ of order n on $(a, b]$ such that $f^{(n)}(x) \in C_\gamma[a, b]$

$$C_\gamma^n[a, b] = \{f : ||f||_{C_\gamma^n} = \Sigma_{k=0}^{n-1} ||f^{(k)}||_C + ||f^{(n)}||_{C_\gamma}\}, \quad C_\gamma^0[a, b] = C_\gamma[a, b].$$

Finally, we have the space of functions for which the solution of (2.14) is well defined in a general sense

$$C_\gamma^{\alpha,r}[a, b] = \{y(x) \in C^r[a, b] :^C D_{a+}^\alpha y \in C_\gamma[a, b]\}, \quad C_\gamma^{r,r}[a, b] = C_\gamma^r[a, b]$$

Finally, we have the following Theorem which states the Global sufficient conditions for existence and uniqueness of the solution to the fractional order differential equation of Caputo-type (2.14)–(2.15).

Theorem 2.2 (Global Existence and Uniqueness) *Let $\alpha > 0$ and $n = -[-\alpha]$, and let $0 \le \gamma \le 1$ and $\gamma \le \alpha$. Let G be an open set in $\mathbb{C}$ and let $f : (a, b] \times G \to \mathbb{C}$ be a function such that for any $y \in G$, $f[x, y] \in C_\gamma[a, b]$ and the Lipschitz condition (2.16) holds.*

- *If $n - 1 < \alpha < n$ ($n \in \mathbb{N}$), then there exists a unique solution $y(x)$ to the Cauchy problem of (2.14)–(2.15) in the space $C_\gamma^{\alpha,n-1}[a, b]$.*
- *If $\alpha = n \in \mathbb{N}$, then there exists a unique solution $y(x)$ to the Cauchy problem (2.18) in the space $C_\gamma^n[a, b]$.*

- *In particular, when $\gamma = 0$ and $f[x, y] \in C[a, b]$, there exist unique solutions to the Cauchy problem (2.14) and (2.15) in the space $C^{\alpha,n-1}[a, b]$:*

$$C^{\alpha,n-1}[a, b] := C_0^{\alpha,n-1}[a, b] = \{y(x) \in C^{n-1}[a, b] :^C D_{a+}^\alpha y \in C[a, b]\},$$
(2.19)

and to the Cauchy problem (2.17) in the space $C^n[a, b]$.

The proof of this theorem follows from the Banach fixed point theorem by means of the reduction of initial value problem (2.14)–(2.15) to its equivalent Volterra integral equation (2.17) (see Theorem 3.25 in [9]).

2.6 Types of Fractional Systems

Throughout the book, two types of fractional systems are handled, then we present the difference between them and the definitions and concepts corresponding to each one. Consider the following fractional order nonlinear system in the form:

$$_{t_0}D_t^{\alpha_i}x_i(t) = f_i(x(t), u), \quad x_i(t_0) = x_{0_i}$$
$$y = h(x, u),$$
(2.20)

for $1 \leq i \leq n$, or in its vector representation

$$_{t_0}D_t^\alpha x = f(x, u), \quad x(t_0) = x_0$$
$$y = h(x, u),$$
(2.21)

where $x \in \mathbb{R}^n$ is the state vector, $f : \mathbb{R}^n \times \mathbb{R}^m \to \mathbb{R}^n$ is a Lipschitz continuous function, $x_0 \in \mathbb{R}^n$ are the initial conditions with $t_0 = 0$ and $y \in \mathbb{R}^p$ denotes the available output (measurable output) of the system, $h : \mathbb{R}^n \times \mathbb{R}^m \to \mathbb{R}^p$ is a continuous function, and $u \in \mathbb{R}^m$ is the vector input and $\alpha = [\alpha_1, \alpha_2, ..., \alpha_n]^T$ for $0 = \alpha_0 < \alpha_i < 1$ ($i = 1, 2, ..., n$). We omit argument t in these notations and write x instead of $x(t)$.

If $\alpha_1 = \alpha_2 = ... = \alpha_n \equiv \alpha$, system (2.20) is called commensurate fractional order system, otherwise it is an incommensurate fractional order system [11]. From now, we consider α for an incommensurate fractional order system and, moreover, $\alpha_1, \alpha_2, ..., \alpha_n$ are not multiples of each other.

In this book, we will work with these two types of fractional systems, each one with different characteristics that it is important to highlight, due to that we are going to study the different properties that have the fractional systems commensurate and incommensurate.

2.7 Algebraic Definitions

2.7.1 Commensurate Fractional Order Systems

Definition 2.4 ([12]) A state variable $x_i \in \mathbb{R}$ of system (2.21) satisfies the Fractional Algebraic Observability (FAO) if x_i is a function of the first $r_1, r_2 \in \mathbb{N}$ sequential fractional derivatives of the available output y and the input u, respectively, i.e.,

$$x_i = \phi_i \left(y, y^{(\alpha)}, \mathscr{D}^{2\alpha} y, \ldots, \mathscr{D}^{r_1\alpha} y, u, u^{(\alpha)}, \mathscr{D}^{2\alpha} u, \ldots, \mathscr{D}^{r_2\alpha} u \right),$$

where $\phi_i : \mathbb{R}^{(r_1+1)p} \times \mathbb{R}^{(r_2+1)m} \to \mathbb{R}$.

From Definition 2.4, we have the following example:

Example 2.3 Classical Chua's oscillator is a simple electronic circuit that exhibits nonlinear dynamical phenomena such as bifurcation and chaos. The model in state equations is given by:

$$\dot{x}_1 = \rho(x_2 - x_1 - f(x))$$
$$\dot{x}_2 = x_1 - x_2 + x_3$$
$$\dot{x}_3 = -\beta x_2 - \gamma x_3,$$

where $f(x) = m_1 x_1 + \frac{1}{2}(m_0 - m_1) \times (|x_1 + 1| - |x_1 - 1|)$, and $\rho, \beta, \gamma, m_0, m_1$ are parameters obtained from the values of the resistances, capacitances, and inductances of the circuit.

The Chua–Hartley's system is different from Chua's system in that the piecewise-linear nonlinearity is replaced by an appropriate cubic nonlinearity which yields very similar behavior. Derivatives on the left side of the differential equations are also replaced by the fractional derivatives [13].

Let the Chua–Hartley fractional commensurate oscillator be given by:

$$D^\alpha x_1 = \rho \left(x_2 + \frac{x_1 - 2x_1^3}{7} \right)$$
$$D^\alpha x_2 = x_1 - x_2 + x_3$$
$$D^\alpha x_3 = -\beta x_2$$

Taking the measurable output as $y = x_3$, the following relations can be obtained:

$$x_1 = \phi_1 \left(y, D^\alpha y, D^{2\alpha} y \right) = -\frac{1}{\beta} D^{2\alpha} y - \frac{1}{\beta} D^\alpha y - y$$

$$x_2 = \phi_2 \left(D^\alpha y \right) = -\frac{1}{\beta} D^\alpha y$$

and, therefore, states x_1 and x_2 satisfy the FAO condition.

For commensurate systems we have the following definition about Picard–Vessiot Systems

Definition 2.5 A fractional order system is fractional Picard–Vessiot (PV) if and only if the vector space generated by the derivatives of the set $\{\mathscr{D}^{\mu\alpha}\overline{y}, \ \mu \in \mathbb{N} \cup \{0\}\}$ has finite dimension, where $\overline{y}$ is the fractional differential primitive element.

Now, consider there exists an element $\overline{y} \in \mathbb{R}$ and let $n \in \mathbb{N} \cup \{0\}$ be the minimum integer such that $\mathscr{D}^{n\alpha}\overline{y}$ is analytically dependent on $\left\{\overline{y}, \overline{y}^{(\alpha)}, \mathscr{D}^{2\alpha}\overline{y}, ..., \mathscr{D}^{[n-1]\alpha}\overline{y}\right\}$, then

$$\mathscr{D}^{n\alpha}\overline{y} = -\mathscr{T}\left(\overline{y}, \overline{y}^{(\alpha)}, \mathscr{D}^{2\alpha}\overline{y}, ..., \mathscr{D}^{[n-1]\alpha}\overline{y}, ...\right.$$
$$\left. ... u, u^{(\alpha)}, \mathscr{D}^{2\alpha}u, ..., \mathscr{D}^{\gamma\alpha}u\right),$$

with $n, \gamma \in \mathbb{N} \cup \{0\}$. Set

$$\eta_j = \mathscr{D}^{((j-1)\alpha)}\overline{y}, \qquad 1 \leq j \leq n, \tag{2.22}$$

then, Fractional Generalized Observability Canonical Form (FGOCF) of system (2.21) is obtained

$$\begin{aligned}
\eta_1^{(\alpha)} &= \eta_2, \\
\eta_2^{(\alpha)} &= \eta_3, \\
&\vdots \\
\eta_{n-1}^{(\alpha)} &= \eta_n, \\
\eta_n^{(\alpha)} &= -\mathscr{T}\left(\eta_1, ..., \eta_n, u, u^{(\alpha)}, \mathscr{D}^{2\alpha}u, ..., \mathscr{D}^{\gamma\alpha}u\right), \\
\overline{y} &= \eta_1.
\end{aligned} \tag{2.23}$$

This yields to the following lemma which is proved as above.

Lemma 2.1 *An observable nonlinear fractional order system* (2.21) *is transformable to a FGOCF if and only if it is PV.* □

There exist some systems that do not necessarily satisfy FAO. Then, the following definition is given.

Definition 2.6 Let $\bar{n}$ states of system (2.21) satisfy FAO property for $\bar{n} < n$, then we will say that system (2.21) is Fractional Liouvillian if the $n - \bar{n}$ states can be obtained by adjunction of fractional order integrals of the $\bar{n}$ states.

As a result from above definition, we can rewrite Definition 2.6 as follows:

Definition 2.7 A state variable $x_i \in \mathbb{R}$ satisfies Fractional Liouvillian Algebraic Observability (FLAO), if x_i is a function of the first $r_1, r_2 \in \mathbb{N}$ sequential fractional derivatives of the available output $\overline{y} = I^\alpha y$ and the input u, respectively, i.e.,

$$x_i = \phi_i \left(I^\alpha y, y, \mathscr{D}^\alpha y, \ldots, \mathscr{D}^{(r_1-1)\alpha} y, \ldots, u, u^{(\alpha)}, \mathscr{D}^{2\alpha} u, \ldots, \mathscr{D}^{r_2\alpha} u \right),$$

where $\phi_i : \mathbb{R}^{(r_1+1)p} \times \mathbb{R}^{(r_2+1)m} \to \mathbb{R}$.

Definition 2.8 A family of systems is Picard–Vessiot (PV), if and only if the vector space generated by the fractional derivatives of the family

$$\left\{ \mathscr{D}^{n_j\alpha} \bar{y}_j, \; n_j \geq 0, \; 1 \leq j \leq q, \; 0 \leq \alpha \leq 1 \right\}$$

has finite dimension, where $\bar{y}_j$ is the $j - th$ output (fractional differential primitive element [14]).

Now, consider next property about diagonalization of companion form matrices, take into account this property for Chaps. 6 and 9:

Property 2.3 ([15]) Given the Hurwitz matrix $A \in \mathbb{R}^{n \times n}$ with different eigenvalues, i.e., $\lambda_i(A) \neq \lambda_j(A)$, there exists a linear transformation $V \in \mathbb{R}^{n \times n}$ such that $D = V^{-1}AV = \text{diag}(\lambda_1, \ldots, \lambda_n)$, where matrix V is the Vandermonde matrix given by

$$V := \begin{pmatrix} 1 & \lambda_1 & \cdots & \lambda_1^{n-1} \\ 1 & \lambda_2 & \cdots & \lambda_2^{n-1} \\ \vdots & \vdots & & \vdots \\ 1 & \lambda_n & \cdots & \lambda_n^{n-1} \end{pmatrix}, \tag{2.24}$$

with the associate monic polynomial $p(\lambda) = \prod_i^n (\lambda - \lambda_i) = p_0 + p_1\lambda + \cdots + \lambda^n$. And its inverse $V^{-1} = HV^T\Delta^{-1}$, where $\Delta = \text{diag}\left(\dot{p}(\lambda_1), \ldots, \dot{p}(\lambda_n) \right)$, $\dot{p}(\lambda_i)$ are valued derivatives of $p(\lambda)$ with respect to $\lambda = \lambda_i$ for $1 \leq i \leq n$ and H is the Hankel matrix given by

$$H = \begin{pmatrix} p_1 & p_2 & \cdots & p_{n-1} & 1 \\ p_2 & p_3 & \cdots & 1 & 0 \\ \vdots & \vdots & & \vdots & \vdots \\ p_{n-1} & 1 & \cdots & 0 & 0 \\ 1 & 0 & \cdots & 0 & 0 \end{pmatrix}.$$

Finally, a generalization of the generalized synchronization definition [12] is defined below for the synchronization of slave families with a master.

Definition 2.9 Let the vectors $X_m = (x_{m_1}, \ldots, x_{n_m}) \in \mathbb{R}^{n_m}$ and $X_s = (x_{s_1}, \ldots, x_{n_s}) \in \mathbb{R}^{n_s}$ be master and slave state vectors families, respectively, then the family of slave systems is in a state of Fractional Generalized Multi-Synchronization (FGMS) with their family of master systems if there exists a family of fractional outputs that generates a transformation $H_{ms} : \mathbb{R}^{n_s} \to \mathbb{R}^{n_m}$ with $H_{ms} = \Phi_m^{-1} \circ \Phi_s$ and there exist an algebraic manifold $M = \{(X_s, X_m) \mid X_m = H_{ms}(X_s)\}$ and a compact set $B \subset \mathbb{R}^{n_m} \times \mathbb{R}^{n_s}$ with $M \subset B$ such that the trajectories with initial conditions in B tend to M as $t \to \infty$. $\qquad \square$

Fig. 2.1 FGS regimen

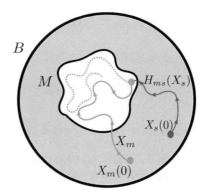

In Fig. 2.1, we can see a better representation of Fractional Generalized Synchronization.

2.7.2 Incommensurate Fractional Order Systems

We introduce the following fractional algebraic observability property for incommensurate fractional order systems.

Definition 2.10 A state variable $x_i \in \mathbb{R}$ satisfies the Incommensurate Fractional Algebraic Observability (IFAO) property if it is a fractional polynomial function of derivatives of the available output y, i.e.,

$$\eta_i = \phi_i(y_{\bar{x}}, y_{\bar{x}}^{(\sum_i^n \alpha_i)}, y_{\bar{x}}^{(\sum_i^n \alpha_i)}, ..., y_{\bar{x}}^{(\sum_i^n \alpha_i)}), 0 \leq \alpha_i \leq 1, \qquad (2.25)$$

where $\phi_i : \mathbb{R}^{(n+1)p} \to \mathbb{R}$.

from Definition 2.10 we have the following example:

Example 2.4 Consider, now, the Chua–Hartley fractional incommensurate oscillator:

$$D^{\alpha_1} x_1 = \rho\left(x_2 + \frac{x_1 - 2x_1^3}{7}\right)$$
$$D^{\alpha_2} x_2 = x_1 - x_2 + x_3$$
$$D^{\alpha_3} x_3 = -\beta x_2$$

Taking the measurable output as $y = x_1$, the following relations can be obtained:

$$x_2 = \phi_1 (y, D^{\alpha_1} y) = \frac{1}{\rho} D^{\alpha_1} y - \frac{1}{7}(y - 2y^3)$$

$$x_3 = \phi_2 (y, D^{\alpha_1} y, D^{\alpha_2} y, D^{\alpha_2} D^{\alpha_1} y) = \frac{1}{\rho} D^{\alpha_2} D^{\alpha_1} y - \frac{1}{7} D^{\alpha_2}(y - 2y^3) - y$$

$$+ \frac{1}{\rho} D^{\alpha_1} y - \frac{1}{7}(y - 2y^3)$$

and, therefore, states x_2 and x_3 satisfy the IFAO condition.

Remark 2.1 Each state that satisfies the FAO (or IFAO) condition is said to be algebraically observable, and thus its dynamics can be reconstructed.

We introduce the concepts of fractional incommensurate differential primitive element and fractional order Picard–Vessiot system (PV).

Definition 2.11 A fractional order system is PV if the vector space generated by the derivatives of the set $\{y^{(\sum_{i=1}^{n} \alpha_i)}, \ n \geq 0, \ 0 \leq \alpha_i \leq 1\}$ has finite dimension where y is the fractional incommensurate differential primitive element.

For introducing the Incommensurate Fractional Generalized Observability Canonical Form (IFGOCF), we have the following:
Consider a element $\bar{y} \in \mathbb{R}$ such that $\bar{y}^{(\sum_{i=1}^{n} \alpha_i)}$ depends on $(\bar{y}, \bar{y}^{(\alpha_1)}, \bar{y}^{(\alpha_1+\alpha_2)}, ..., \bar{y}^{(\sum_{i=1}^{n-1} \alpha_i)})$:

$$\bar{H}(\bar{y}, \bar{y}^{(\alpha_1)}, \bar{y}^{(\alpha_1+\alpha_2)}, \ldots, \bar{y}^{(\sum_{i=1}^{n} \alpha_i)},$$
$$u, u^{(\alpha_1)}, u^{(\alpha_1+\alpha_2)}, \ldots, u^{(\sum_{i=1}^{n} \alpha_i)}) = 0 \tag{2.26}$$

The system (2.21) can be solved as

$$\bar{y}^{(\sum_{i=1}^{n} \alpha_i)} = -\mathscr{L}(\bar{y}, \bar{y}^{(\alpha_1)}, \bar{y}^{(\alpha_1+\alpha_2)}, ..., \bar{y}^{(\sum_{i=1}^{n-1} \alpha_i)}, u,$$
$$u^{(\alpha_1)}, u^{(\alpha_1+\alpha_2)}, ..., u^{(\sum_{i=1}^{n-1} \alpha_i)}) + u^{(\sum_{i=1}^{n} \alpha_i)}$$

Redefining $z_j = \bar{y}^{(\sum_{i=0}^{j-1} \alpha_i)}$, for $1 \leq j \leq n$, we introduced the IFGOCF of (2.21) as

$$z_1^{(\alpha_1)} = z_2$$
$$z_2^{(\alpha_2)} = z_3$$
$$\vdots$$
$$z_{n-1}^{(\alpha_{n-1})} = z_n$$
$$z_n^{(\alpha_n)} = -\mathscr{L}(z_1, z_2, z_3, ..., z_{n-1}, u, u^{(\alpha_1)},$$
$$u^{(\alpha_1+\alpha_2)}, ..., u^{(\sum_{i=1}^{n-1} \alpha_i)}) + u^{(\sum_{i=1}^{n} \alpha_i)}$$
$$\bar{y} = z_1 \tag{2.27}$$

Then, the next result is immediate from the above and from the Definition 2.11, we establish the following lemma for incommensurate systems.

Lemma 2.2 *A nonlinear system of fractional order incommensurate* (2.21) *is transformable to the IFGOCF:*

$$z_{ij}^{(\alpha_{ij})} = z_{ij+1}, \quad 1 \le j \le n-1 \tag{2.28}$$
$$z_{in}^{(\alpha_{in})} = -\mathscr{L}_i(z_i, u_i, ..., D^{\alpha_{in-1}} \cdots D^{\alpha_{i1}} u_i) + D^{\alpha_{in}} \cdots D^{\alpha_{i1}} u_i$$
$$\bar{y}_i = z_{i1}$$

if and only if is PV, where $\bar{y}_i$ is the incommensurate fractional primitive element. $\square$

Remark 2.2 It is normally expected that in the presence of parameter mismatches and noise, synchronization error is not asymptotically stable in the integer order case and commensurate fractional order cases [16–19]. Here, even in the absence of latter elements, synchronization is not perfectly achieved for the case of incommensurate fractional order systems in a generalized sense.

We introduce a new definition of Fractional Generalized quasi-Synchronization (FGqS) which depends on the existence of the fractional incommensurate differential primitive element.

Definition 2.12 Let $t_0 \ge 0$ be the initial time. The slave (9.3) and master systems (9.2) are said to be in a state of $FGqS$ with bound ϵ if there exist $T := T(\epsilon) \ge t_0$ and outputs that generate a transformation $H_{MS} : \mathbb{R}^{n_S} \to \mathbb{R}^{n_S}$ with $H_{MS} = \phi_M^{-1} \circ \phi_S$ as well as there exists a quasi-Synchronization algebraic manifold $M_q = \{(x_S, x_M) | \|x_M - H_{MS}(x_S)\| \le \epsilon\}$ and a compact set $B \subset \mathbb{R}^{n_S} \times \mathbb{R}^{n_M}$ with $M_q \subset B$ such that their trajectories with initial conditions in B are in M_q for $t \ge T$.

Figure 2.2 make things intuitively clear.

Fig. 2.2 FGqS regimen

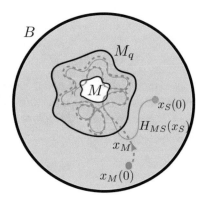

Fig. 2.3 Matignon stability

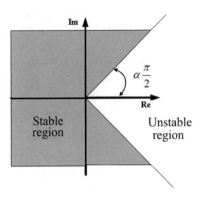

2.8   Stability Results Commensurate Systems

Theorem 2.3 (Matignon [20]) *Let* $\alpha < 2$ *and* $\bar{A} \in \mathbb{C}^{n \times n}$. *The autonomous system*

$$x^{(\alpha)} = \bar{A}x \quad with \quad x(0) = x_0$$

is asymptotically stable if and only if $|arg(\lambda_i(\bar{A}))| > \alpha\pi/2$, *where* $\lambda_i(\bar{A})$ *is the i-th eigenvalue of the matrix* $\bar{A}$ (Fig. 2.3). □

Remark 2.3 As a particular case from above Theorem, for $0 < \alpha < 1$, all Hurwitz matrix satisfy the condition

$$|arg(\lambda_i(A))| > \frac{\pi}{2} > \frac{\alpha\pi}{2}$$

Definition 2.13 The Kronecker product of matrices $A \in \mathbb{R}^{m \times n}$ and $B \in \mathbb{R}^{p \times q}$ is defined as

$$A \otimes B = \begin{pmatrix} a_{11}B & \cdots & a_{1n}B \\ \vdots & \ddots & \vdots \\ a_{m1}B & \cdots & a_{mn}B \end{pmatrix} \in \mathbb{R}^{mp \times nq}.$$

Theorem 2.4 ([4]) *Let* $x = 0$ *be an equilibrium point for the system* $D^{\alpha}x(t) = f(t, x)$, *and* $\mathbb{D} \subset \mathbb{R}^n$ *be a domain containing the origin. Let* $V(t, x(t)) : [0, \infty) \times \mathbb{D} \to \mathbb{R}$ *be a continuously differentiable function and locally Lipschitz with respect to* x *such that*

$$\alpha_1 \|x\|^a \leq V(t, x(t)) \leq \alpha_2 \|x\|^{ab}$$
$$D^{\beta}V(t, x(t)) \leq -\alpha_3 \|x\|^{ab},$$

where $t \geq 0$, $x \in \mathbb{D}$, $\beta \in (0, 1)$, α_1, α_2, α_3, a and b arbitrary positive constants. Then, $x = 0$ is Mittag–Leffler stable. If the assumptions hold globally on $\mathbb{R}^n$, then $x = 0$ is globally Mittag–Leffler stable.

Lemma 2.3 ([21]) *Let $x(t) \in \mathbb{R}$ be a continuous and derivable function. Then, for any time instant $t \geq t_0$*

$$\frac{1}{2} D^{\alpha} x^2(t) \leq x(t) D^{\alpha} x(t) \qquad \forall \alpha \in (0, 1)$$

2.8.1   Gronwall–Bellman Generalized Lemma

In 1918, Gronwall gave the Gronwall–Bellman inequality (see [22]), which is an important tool in the study of boundedness, uniqueness and other aspects of qualitative behavior of solutions of differential equations and stability. Many authors gave a significant number of generalizations about this inequality (see [23, 24]) being important for applications in differential and integral equations of integer and fractional order. In this section, we will highlight the fractional formulation and the usefulness of this tool for stabilization of fractional input-affine systems. In fractional nonlinear systems, there is very little theory about the stabilization problem through the Gronwall–Bellman Lemma. In 2011, N'Doye showed some results for the stabilization of a particular class of nonlinear system, certain technical problems have been found in the demonstration, in this section, we try to give a correction to those problems.

Similar to the exponential function frequently used in the solutions of integer order, a function frequently used in the solutions of fractional order is the Mittag–Leffler function with two parameters defined as

$$E_{\alpha, \beta}(z) = \sum_{k=0}^{\infty} \frac{z^k}{\Gamma(\alpha k + \beta)}, \quad \alpha, \beta > 0$$

with the gamma function $\Gamma(z)$ defined by

$$\Gamma(z) = \int_0^{\infty} e^{-t} t^{z-1} dt, z \in \mathbb{C} \tag{2.29}$$

Now, we mention some results that will be very useful to demonstrate the stability of a type of fractional systems.

Corollary 2.1 *If $A \in \mathbb{C}^{n \times n}$ and $0 < \alpha < 2$, β is an arbitrary real number, η is such that $\alpha \pi / 2 < \eta < \min\{\pi, \pi \alpha\}$ and $\theta \geq 0$ is a real constant, then*

$$\|E_{\alpha, \beta}(A)\| \leq \frac{\theta}{1 + \|A\|}, \quad \eta \leq arg(\lambda_i(A)) \leq \pi, \ 1 \leq i \leq n, \tag{2.30}$$

where $\lambda(A)$ denotes the eigenvalues of the matrix A and $|| \cdot ||$ denotes the $l_2 - norm$.

For the demonstration of exponential stability theorems, the following generalizations of the Gronwall–Bellman Lemma will be very useful. The following result will be used mainly for integer systems and the proof can be found in [13].

Lemma 2.4 ([13]) *Let the functions ψ, φ, b and k be continuous and non-negative of $J = [\alpha, \beta]$, n be a positive, $n \geq 2$ and φ/b be a nondecreasing function. If*

$$\psi(t) \leq \varphi(t) + b(t) \int_\alpha^t k(s)\psi^n(s)ds, \ t \in J, \tag{2.31}$$

then,

$$\psi(t) \leq \varphi(t)\{1 - (n-1)\int_\alpha^t k(s)b(s)\varphi^{n-1}(s)ds\}^{\dfrac{1}{n-1}}$$

$$\alpha \leq t \leq \beta_n, \ \beta_n = \sup\{t \in J : (n-1)\int_\alpha^t kb\varphi^{n-1}ds < 1\} \tag{2.32}$$

For the case of fractional input-affine systems is necessary to use a generalization slightly different from the traditional Gronwall–Bellman Lemma in [24], the authors proved the following result:

Theorem 2.5 ([24]) *Let $\psi(t)$ be a continuous, and nonnegative function, $k(t, s)$ be a continuously differential function in t and continuous in $s, k(t, s) \geq 0$ for $t \geq s \geq t_0$. If*

$$\psi(t) \leq c + \int_{t_0}^t k(t, s)\psi^n(s)ds, \ c > 0 \tag{2.33}$$

Then,

$$\psi(t) \leq \left[c^{-(n-1)} - (n-1)\{\int_0^t \left[k(t, t) + \int_0^s \frac{\partial}{\partial s}k(s, r)dr\right]ds\}\right]^{\dfrac{1}{1-n}} \tag{2.34}$$

The following result will be used to show the stability in the case of controlled systems by means of a dynamical control.

Lemma 2.5 ([13]) *Let the positive continuous functions ψ, φ, b and k in $J = [\alpha, \beta], n, m \in \mathbb{N}$ such that $n, m \geq 2$ and a nondecreasing function $\varphi(t)$. If,*

$$\psi(t) \leq \varphi(t) + \int_\alpha^t k_1(s)\psi^n(s) + k_2\psi^m(s)ds, \ t \in J,$$

then

$$\psi(t) \le \varphi(t) \left[1 - (n + m - 2) \int_\alpha^t k_1(s)\varphi^{n-1}(s) + k_2(s)\varphi^{m-1}(s)ds \right]^{\dfrac{1}{n+m-2}}, \ \alpha \le t \le \beta_n,$$

where

$$\beta_n = sup\{t \in J : (n + m - 2) \int_\alpha^t k_1\varphi^{n-1} + k_2\varphi^{m-1}ds < 1\}$$

2.8.1.1   Stabilization of Integer Input-Affine Systems

This section is divided into two parts; in the first part, we present the conditions that guarantee exponential stability for systems of the form (2.35), controlled by means of a static state feedback, and in the second part, necessary conditions and results are shown for systems controlled by means of a dynamical control. Let integer input-affine systems of the form

$$\dot{x}(t) = Ax(t) + \sum_{i=1}^{n} g_i(x(t))u_i(t) + Bu(t)$$

$$y(t) = Cx(t), \quad x(0) = x_0, \tag{2.35}$$

where A, B, C, are known constant matrices, $x \in \mathbb{R}^n$ is the state vector, $y \in \mathbb{R}^p$ is the output vector and $u \in \mathbb{R}^m$ is the input vector.

2.8.1.2   Static State Feedback Control

The following theorem, gives conditions for the exponential stability of the systems of the form above given.

Theorem 2.6 ([13]) *Consider the input-affine system given by (2.35), that satisfy the following conditions*

- *For $1 \le i \le m$, there exists an integer $q \ge 1$, such that*

$$||g_i(x(t))|| \le \mu_i ||x(t)||^q, \tag{2.36}$$

where μ_i are positive constant and $g_i(0) = 0$ and we define $\mu = \sum_{i=1}^{m} \mu_i$.
- *The pair (A, B) is stabilizable.*

The system (2.35) controlled by means of a static state feedback, $u(t) = Kx(t)$, is exponentially stable if all eigenvalues of matrix $(A + BK)$ have a strictly negative real part and if

$$0 < ||x_0|| \leq \epsilon_0, \quad \epsilon_0^q < \frac{|\omega|}{\mu ||L|| M^{q+1}}, \tag{2.37}$$

where $M > 0$ y $\omega < 0$ are scalars that satisfy

$$||exp((A + BK)t)|| < M exp(\omega t), \quad \forall t \geq 0.$$

Moreover, the state $x(t)$ is bounded as follows

$$||x(t)|| \leq \frac{\epsilon_0 M exp(\omega t) ||x_0||}{\left(1 - \dfrac{M^{q+1}\mu ||L|| \epsilon_0^q ||x_0||^q}{|\omega|}\right)^{\frac{1}{q}}} \tag{2.38}$$

The proof is given in [13].   □

We show a numerical example to illustrate the previous result (see the Appendix A.1).

2.8.1.3 Dynamical Control

Any form given by an integer input-affine system, (2.35) can be carried out to the canonical form (see [25])

$$\dot{z}_1 = z_2$$
$$\dot{z}_2 = z_3$$
$$\vdots$$
$$\dot{z}_n = \mathscr{F}(z, u, \dot{u}, \ldots, u^{(n-1)}) + \bar{u} \tag{2.39}$$
$$u_1 = u$$
$$\dot{u}_1 = u_2$$
$$\dot{u}_2 = u_3$$
$$\vdots$$
$$\dot{u}_n = \bar{u} = -\mathscr{F}(z, u_1, u_2, \ldots, u_n) = \bar{K}z,$$

where the control, $\bar{u}$ is above defined and the gain vector $\bar{K}$ is such that the matrix

$$\begin{pmatrix} 0 & 1 & 0 & 0 \\ \vdots & \vdots & \ddots & \vdots \\ 0 & \ldots & 0 & 1 \\ -k_1 & -k_2 & \ldots & -k_n \end{pmatrix} \tag{2.40}$$

has its eigenvalues with negative real part.

The original system can be controlled by means of applying a dynamical control u that is to say

$$\dot{x} = Ax + \sum_{i=1}^{n} g(x)_i u_i + Bu \tag{2.41}$$

The following theorem establishes sufficient conditions to guarantee the exponential stability of the trajectories for the case of integer input-affine systems given by (2.41).

Theorem 2.7 *Let integer input-affine system satisfying*

- *The nonlinear part of $g_i(x)$ is bounded by*

$$||g_i(x)|| \leq \mu_i ||x||^q \tag{2.42}$$

- *The control u is bounded by $||u|| \leq \rho ||x||^p$ where $p \geq 1$ and at least their n first derivatives are bounded.*
- *The matrix A has all its eigenvalues with negative real part.*

They are exponentially stable systems if there exists a K such that the matrix given in (2.40) has its eigenvalues with negative real part and the initial condition of the system (2.42) satisfies the following inequality.

$$0 < ||\bar{x}_0|| \leq \epsilon_0 \tag{2.43}$$

for ε_0^{2p+q-2} such that $v(\varepsilon_0) > 1$, where

$$v(\varepsilon_0) = (2p+q-2)\left[(M\varepsilon_0)^{p-1}\int_0^t MB\rho exp((p-1)\tau\omega)d\tau\right]$$
$$+ (2p+q-2)\left[(M\varepsilon_0)^{q+p-1}\int_0^t M\mu\rho exp((q+p-1)\tau\omega)d\tau\right] \tag{2.44}$$

The constants $M > 0$ and $\omega < 0$ are given by

$$||exp((A)t)|| < Mexp(\omega t), \quad \forall t \geq 0 \tag{2.45}$$

In addition, the state norm, $x(t)$, is bounded, i.e.,

$$||x(t)|| \leq \frac{\varepsilon_0 Mexp(\omega t)\varepsilon_0}{\left(1 - \dfrac{M^{q+1}\mu\varepsilon_0^q\varepsilon_0^{q-1}}{|\omega|}\right)^{\frac{1}{q-1}}} \tag{2.46}$$

Proof The solution of (2.41) is given by

$$x(t) = exp(A(t))x_0 + \int_0^t exp(A(t - \tau))u(\tau)[B + G(x)]d\tau$$

applying the norm on both sides of equality, we have

$$||x(t)|| \leq Mexp(\omega t)\varepsilon_0 + \int_0^t Mexp(\omega t)exp(-\omega t)\left[||B||\rho||x|| + \mu\rho||x||^{q+p}\right]d\tau$$
$$(2.47)$$

this yields to

$$||x||exp(-\omega t) \leq M\varepsilon_0 + MB\rho \int_0^t exp((p - 1)\omega\tau)||x||^p exp(-p\omega\tau)d\tau + \quad (2.48)$$
$$+M\mu\rho||u|| \int_0^t exp(q + p - 1)\omega\tau exp(-(q + p)\omega\tau)||x||^q d\tau$$

Finally, applying Lemma 2.5 for the functions

$$\psi(t) = ||x||exp(-\omega\tau)$$
$$\varphi(t) = M\varepsilon_0$$
$$k_1(t) = MB\rho\mu exp((p - 1)\omega\tau)$$
$$k_2(t) = M\mu\rho\mu exp((q + p - 1)\omega\tau)$$

and since ε_0 satisfies the inequality (2.42), we have

$$||x|| \leq \frac{\varepsilon_0 exp(\omega t)\varepsilon_0}{(1 - v(\varepsilon_0))^{1/(2p+q-1)}} \quad (2.49)$$

□

To illustrate the results of the Dynamical Control (see the Appendix A.2).

2.8.2 *Stabilization of Fractional Input-Affine Systems*

In this section, stabilization is discussed for fractional input-affine systems of the form (2.50), this section is divided in two parts, the first gives conditions for systems controlled by a static state feedback, finally, in the second part, we give conditions for systems controlled by means of a dynamical control.

The fractional input-affine systems, with derivation order $0 < \alpha < 1$, are of the form

$$D^\alpha x(t) = Ax(t) + \sum_{i=1}^{n} g_i(x(t))u_i(t) + Bu(t) \tag{2.50}$$

$$y(t) = Cx(t), \quad x(0) = x_0,$$

where A, B, C are constant matrix of appropriate size.

2.8.2.1 Static Feedback Control

The following theorem establishes conditions for the exponential stability of fractional input-affine systems of the form (2.50). Here, we make a modification in the proof previously proposed by N'Doye.

Theorem 2.8 *Suppose the system* (2.50) *satisfies*

- *For all $1 \le i \le m$, there exists an integer $q \ge 1$, such that*

$$||g_i(x(t))|| \le \mu_i ||x(t)||^q, \tag{2.51}$$

where μ_i are positive constants and $g_i(0) = 0$. As in the integer case, $\mu = \sum_{i=1}^{m} \mu_i$.
- *The pair (A,B) is stabilizable*

Then, the system (2.50) *controlled by means of a static-state feedback, $u(t) = Kx(t)$ is asymptotically stable if the matrix $W = A + BK$, satisfies $|arg(\lambda(W))| > \alpha \dfrac{\pi}{2}$ and the initial condition satisfies*

$$||x_0|| < \left(\frac{\alpha ||W||}{2\mu ||L|| \theta^{q+1}} \right)^{\frac{1}{q}} \tag{2.52}$$

In addition, the solution is bounded

$$||x(t)|| \le \frac{\dfrac{\theta ||x_0||}{1 + ||W||t^\alpha}}{\left(1 - \dfrac{2\mu\theta^{q+1}||L||||x_0||^q}{\alpha ||W||} \left(1 - \dfrac{1}{1 + ||W||(t/2)}\right)\right)^{\frac{1}{q}}}, \tag{2.53}$$

where the constant θ is given by the Corollary 2.1.

Proof Using the control law, $u(t) = Kx(t)$, and applying the Laplace transform to the system (2.50), with $C = I_n$, we have

$$X(s) = (I_n s^\alpha - W)^{-1} \left(s^{\alpha-1} x_0 + \mathscr{L} \left(\sum_{i=1}^{m} g_i(x(t))(Lx(t))_i \right) \right),$$ (2.54)

where $W = (A + BL)$. By hypothesis there exists K such that $|arg(\lambda_i(W))| > \alpha \dfrac{\pi}{2}$. Now, applying the inverse Laplace transform to equation (2.54) we have

$$x(t) = E_{\alpha,1}(Wt^\alpha)x_0 + \int_0^t \left[(t-\tau)^{\alpha-1} E_{\alpha,\alpha}(W(t-\tau)^\alpha) \left(\sum_{i=1}^{m} g_i(x(\tau))(Lx(\tau))_i \right) \right] d\tau$$ (2.55)

Applying the norm on both sides of the relationsship (2.55) and using Corollary 2.1, we have the following inequality:

$$||x(t)|| \le \frac{\theta||x_0||}{1 + ||Wt^\alpha||} + \int_0^t \frac{\mu\theta||L||||t-\tau||^{\alpha-1}}{1 + ||W(t-\tau)^\alpha||} ||x(\tau)||^{q+1} d\tau$$ (2.56)

Rewriting the previous inequality,

$$||x(t)|| \le \frac{\theta||x_0||}{1 + ||W||t^\alpha} + \int_0^t \frac{\mu\theta||L||t-\tau^{\alpha-1}}{1 + ||W(t-\tau)^\alpha||} ||x(\tau)||^{q+1} d\tau$$ (2.57)

It is clear that the term $\dfrac{\theta||x_0||}{1 + ||W||t^\alpha}$ of the previous inequality is bounded by a positive constant, let say R, hence

$$||x(t)|| \le R + \int_0^t \frac{\mu\theta||L||t-\tau^{\alpha-1}}{1 + ||W(t-\tau)^\alpha||} ||x(\tau)||^{q+1} d\tau$$ (2.58)

Now, it is possible to apply the Theorem 2.5 to obtain the following inequality

$$||x(t)|| \le \left[R^{-(n-1)} - (n-1) \left(\int_0^t \left[k(t,t) + \int_0^s \frac{\partial}{\partial s} k(s,r) dr \right] dr \right) \right]^{1/(n-1)}$$ (2.59)

with

$$k(t,\tau) = \frac{\mu\theta||L||(t-\tau)^{\alpha-1}}{1 + ||W(t-\tau)^\alpha||} \qquad n = q+1 \qquad c = R$$

It should be noted that N'Doye omits the fact that the k function depends on two variables, so the generalization used is not appropriate. Hence, we have a modification that is showed in the inequality above.

Given the form of the function $k(t, \tau)$, we have $k(t, t) = 0$ and

$$\int_0^s \frac{\partial}{\partial s} k(s, r) dr = k(s, r)$$

replacing in (2.59), we have

$$\|x(t)\| \leq \left[R^{-(n-1)} - (n-1) \left\{ \int_0^t k(t, \tau) d\tau \right\} \right]^{1/(n-1)}$$

We show a numerical example to illustrate the previous result (see the Appendix B.1).

2.8.2.2 Fractional Dynamical Control

Now, for fractional input-affine systems of the form (2.50), it is possible to transform the system to the fractional canonical form

$$
\begin{aligned}
D^\alpha z_1 &= z_2 \\
D^\alpha z_2 &= z_3 \\
&\vdots \\
D^\alpha z_n &= \mathscr{F}(x, u) + \bar{u} \\
u_1 &= u \\
D^\alpha u_1 &= u_2 \\
D^\alpha u_2 &= u_3 \\
&\vdots \\
D^\alpha u_n &= \bar{u} = -\mathscr{F}(z, u_1, u_2, ..., u_n) = -\bar{K} z
\end{aligned}
\tag{2.60}
$$

with gain vector $\bar{K}$ and

$$
\kappa = \begin{bmatrix}
0 & 1 & 0 & 0 \\
\vdots & \vdots & \ddots & \vdots \\
0 & \cdots & 0 & 1 \\
-k_1 & -k_2 & \cdots & -k_n,
\end{bmatrix}
\tag{2.61}
$$

where the eigenvalues satisfy $|arg(\lambda_i(\kappa))| > \alpha\dfrac{\pi}{2}$. By replacing the fractional dynamic control in the original system, we have

$$
\begin{aligned}
D^\alpha x &= Ax + G(x)u + Bu \\
u_1 &= u \\
D^\alpha u_1 &= u_2 \\
D^\alpha u_2 &= u_3 \\
&\vdots \\
D^\alpha u_n &= \bar{u} = -\psi(x, u_1, u_2, \dots, u_n) + \bar{K}x
\end{aligned}
\tag{2.62}
$$

Now, the following theorem establishes sufficient conditions to guarantee stability of the trajectories for the case of fractional input-affine systems.

Theorem 2.9 *Suppose that* (2.62) *satisfies:*

- *For all* $1 \le i \le m$, *there exist an integer* $q \ge 1$ *such that*

$$
\|g_i(x(t))\| \le \mu_i \|x(t)\|^q,
\tag{2.63}
$$

  where μ_i *are positive constants and* $g_i(0) = 0$. *As in the integer case* $\mu = \sum_{i=1}^{m} \mu_i$.
- *The matrix* A *is such that* $|arg(\lambda_i(A))| > \alpha\dfrac{\pi}{2}$

- *The pair* (A, B) *is stabilizable*

- *The control,* $u(t)$ *is bounded by* $\rho\|x\|^p$ *and at least its first* n *derivatives are bounded.*

The system is stable if there exists $\bar{K}$ *such that* (2.60) *is stable and the initial condition satisfies* $\|x_0\| < \varepsilon$, *where the constant* ε_0 *is such that the following inequality is satisfied for all* $t > 0$

$$
\begin{aligned}
v(t, \varepsilon_0) = 1 - (q + p - 2) \int_0^t \left(\frac{\theta\varepsilon_0}{1 + \|A\|t^\alpha} \right)^{q+p-1} \frac{\rho\mu\theta(t-\tau)^{\alpha-1}}{1 + \|A\|(t-\tau)^\alpha} \\
- (q + p - 2) \int_0^t \left(\frac{\theta\varepsilon_0}{1 + \|A\|t^\alpha} \right)^{p-1} \frac{B\rho\theta(t-\tau)^{\alpha-1}}{1 + \|A\|(t-\tau)^\alpha} > 0
\end{aligned}
\tag{2.64}
$$

The norm is bounded by

$$
\|x(t)\| \le \frac{\dfrac{\theta\varepsilon_0}{1 + \|A\|t^\alpha}}{v(t, \varepsilon_0)^{1/(2p+q-2)}}
\tag{2.65}
$$

the constant θ is given by Corollary 2.1.

Proof First, we apply the Laplace transform to the system (2.62), and we have

$$X(s) = (I_n s^\alpha - A)^{-1} \left(s^{\alpha-1} x_0 + \mathcal{L}(G(x(t))u + Bu) \right) \qquad (2.66)$$

By hypothesis $|arg(\lambda_i(A))| > \alpha \dfrac{\pi}{2}$. Applying the inverse Laplace transform to the Eq. (2.66), we have,

$$x(t) = E_{\alpha,1}(At^\alpha x_0 + \int_0^t \left[(t-\tau)^{\alpha-1} E_{\alpha,\alpha}(A(t-\tau)^\alpha)(G(x(t))u + Bu) \right] d\tau \qquad (2.67)$$

Applying the norm on both sides of the previous equality and using Corollary 2.1, we obtain

$$||x(t)|| \le \frac{\theta||x_0||}{1 + ||At^\alpha||} || \int_0^t \frac{\mu\theta||t-\tau||^{\alpha-1}}{1 + ||A(t-\tau)^\alpha||} ||x(\tau)||^q d\tau + \int_0^t \frac{B\rho\theta||t-\tau||^{\alpha-1}}{1 + ||A(t-\tau)^\alpha||} ||x(\tau)||^p d\tau \qquad (2.68)$$

Now, rewriting the terms, we get

$$||x(t)|| \le \frac{\theta||x_0||}{1 + ||A||t^\alpha} + \int_0^t \frac{\mu\theta(t-\tau)^{\alpha-1}}{1 + ||A(t-\tau)^\alpha||} ||x(\tau)||^q d\tau + \int_0^t \frac{B\rho\theta(t-\tau)^{\alpha-1}}{1 + ||A(t-\tau)^\alpha||} ||x(\tau)||^p d\tau \qquad (2.69)$$

as well as $\dfrac{\theta||x_0||}{1 + ||A||t^\alpha} \le R$, thus

$$||x(t)|| \le R + \int_0^t \frac{\mu\theta||L||(t-\tau)^{\alpha-1}}{1 + ||W(t-\tau)^\alpha||} ||x(\tau)||^q d\tau + \int_0^t \frac{B\rho\theta(t-\tau)^{\alpha-1}}{1 + ||A(t-\tau)^\alpha||} ||x(\tau)||^p d\tau \qquad (2.70)$$

Now, it is possible to apply the Theorem 2.6, with

$$k_1(t,\tau) = \frac{\mu\theta||L||(t-\tau)^{\alpha-1}}{1 + ||A(t-\tau)^\alpha||}$$

$$k_2(t,\tau) = \frac{B\rho\theta||L||(t-\tau)^{\alpha-1}}{1 + ||A(t-\tau)^\alpha||}, \quad c = R$$

This yields to $||x(t)|| \le \dfrac{\dfrac{\theta\varepsilon}{1 + ||A||t^\alpha}}{v(t, \varepsilon_0)^{1/(2p+q-2)}}$

To illustrate the results of the Fractional Dynamical control case (see the Appendix B.2).

2.9   Stability Result Incommensurate Systems

Consider the following stability result for fractional incommesurate order lineal system

$$\frac{D^{(\alpha_1)}x_1(t)}{dt^{(\alpha_1)}} = a_{11}x_1 + a_{12}x_2 + \ldots + a_{1n}x_n$$

$$\frac{D^{(\alpha_2)}x_2(t)}{dt^{(\alpha_2)}} = a_{21}x_1 + a_{22}x_2 + \ldots + a_{2n}x_n \qquad (2.71)$$

$$\vdots$$

$$\frac{D^{(\alpha_n)}x_n(t)}{dt^{(\alpha_n)}} = a_{n1}x_1 + a_{n2}x_2 + \ldots + a_{nn}x_n,$$

where α_i is a rational number between 0 and 1. Let M be the lowest common multiple of the denominators $u_i s$ of $\alpha_i s$, where $\alpha_i = v_i/u_i$, $(u_i, v_i) = 1, u_i, v_i \in Z^+$, $i = 1, ..., n$ and set $\gamma = 1/M$. Then the zero solution of system (2.71) is Lyapunov globally asymptotically stable if all the roots λ_i's of the equation.

$$det \begin{bmatrix} \lambda^{M\alpha_1} - a_{11} & -a_{12} & \cdots & -a_{1_n} \\ -a_{21} & \lambda^{M\alpha_2} - a_{22} & \cdots & -a_{2_n} \\ \vdots & \vdots & \ddots & \vdots \\ -a_{n1} & -a_{n2} & \cdots & \lambda^{M\alpha_n} - a_{nn} \end{bmatrix} = 0 \qquad (2.72)$$

satisfies $|arg(\lambda)| > \gamma\alpha/2$.

References

1. Podlubny, Igor, Fractional differential equations: an introduction to fractional derivatives, fractional differential equations, to methods of their solution and some of their applications, Vol. 198. Elsevier, (1998).
2. Humbert, P., and R. P. Agarwal, Sur la fonction de Mittag-Leffler et quelques-unes de ses généralisations, Bull. Sci. Math 77.2 pp. 180–185, (1953).
3. Luchko, Yuri, and Rudolf Gorenflo, The initial value problem for some fractional differential equations with the Caputo derivatives, (1998).
4. Y. Li, Y. Chen, I. Podlubny, Stability of fractional-order nonlinear dynamic systems: Lyapunov direct method and generalized Mittag-Leffler stability, Computers and Mathematics with Applications 59(5) (2010) 1810–1821.
5. I. Podlubny, Fractional Differential Equations: An Introduction to Fractional Derivatives, Fractional Differential Equations, to Methods of Their Solution and Some of Their Applications, first ed., Academic Press, 1999.
6. I. Podlubny, Geometric and physical interpretation of fractional integration and fractional differentiation, Fractional Calculus and Applied Analysis 5, 367–386, (2002).
7. K. S. Miller and B. Ross, *An introduction to the fractional calculus and fractional differential equations*, New York: John Wiley and Sons Inc., 1993.

8. K. B. Oldham and J. Spanier, *The fractional calculus: Theory and applications of differentiation and integration of arbitrary order*, Dover Books on Mathematics, 2006.
9. A. A. Kilbas, H. M. Srivastava and J. J. Trujillo, *Theory and applications of fractional differential equations*, Elsevier B.V., 2006.
10. Caputo, Michele, Linear models of dissipation whose Q is almost frequency independent II, Geophysical Journal International 13.5 pp. 529–539, (1967).
11. Caponetto, Riccardo, et al. Fractional order systems: modeling and control applications. 2010.
12. Martínez-Guerra, Rafael, and Juan L. Mata-Machuca, Fractional generalized synchronization in a class of nonlinear fractional order systems, Nonlinear Dynamics 77.4, pp. 1237–1244, (2014).
13. Ibrahima N'Doye. Generalisation du lemme de Gronwall-Bellman pour la stabilisation des systemes fractionnaires. PhD thesis, Universite Henri Poincare-Nancy I; Universite Hassan II Ain Chock de Casablanca, 2011.
14. Kolchin, E.R, Differential Algebra and Algebraic Groups, Academic, New York (1973)
15. U. Luther and K. Rost: Matrix exponentials and inversion of confluent Vandermonde matrices. Electronic Transactions on Numerical Analysis, 18, 9–100 (2004)
16. Liu, Xiwei, Ying Liu, and Lingjun Zhou, Quasi-synchronization of nonlinear coupled chaotic systems via aperiodically intermittent pinning control, Neurocomputing 173, pp. 759–767, 2016.
17. He, Wangli, et al, Quasi-synchronization of heterogeneous dynamic networks via distributed impulsive control: error estimation, optimization and design, Automatica 62, pp. 249–262, 2015.
18. Vincent, U. E., et al, Quasi-synchronization dynamics of coupled and driven plasma oscillators, Chaos, Solitons & Fractals 70, pp. 85–94, 2015.
19. Yang, Xujun, et al, Quasi-uniform synchronization of fractional-order memristor-based neural networks with delay, Neurocomputing 234, pp. 205–215, 2015.
20. Matignon, D. Stability results for fractional differential equations with applications to control processing. In Computational engineering in systems applications, Vol. 2, pp. 963–968, 1996.
21. N. Aguila-Camacho, M.A. Duarte-Mermoud, J.A. Gallegos, Lyapunov functions for fractional order systems, Communications in Nonlinear Science and Numerical Simulation 19 (2014) 2951–2957.
22. Thomas Hakon Gronwall. Note on the derivatives with respect to a parameter of the solutions of a system of differential equations. Annals of Mathematics, pp. 292–296, 1919
23. DS Mitrinovic, JE Pecaric, and AM Fink. Inequalities of Gronwall type of a single variable. In Inequalities Involving Functions and Their Integrals and Derivatives, pages 353400. Springer, 1991.
24. BG Pachpatte. Inequalities applicable to retarded Volterra integral equations. Tamkang Journal of Mathematics, 35(4):285–292, 2004
25. Rafael Martínez-Guerra and Christopher Diego Cruz-Ancona. Algorithms of Estimation for Nonlinear Systems: A Differential and Algebraic Viewpoint. Springer, 2017.

Chapter 3
Synchronization of Chaotic Systems by Means of a Nonlinear Observer: An Application to Secure Communications

3.1 Introduction

Applications of the synchronization phenomenon of dynamical systems are important in several applications as in technology and have a wealth of science. Some applications of master–slave synchronization include the control of chaos and chaotic signal masking, where several methodologies have been considered [1, 2]. Nonetheless, it has been established that synchronization of chaotic dynamical systems is not only possible but it is also believed to have potential applications in communication [3, 4]. The strategy is that when is transmitted a message; it is possible to mask it with louder chaos. An outside listener only detects the chaos, which signals like meaningless noise. But if the receiver has an adequate synchronization algorithm that perfectly reproduces the chaos, the receiver can subtract the chaotic mask and detect the original message. Chaos, the receiver can subtract the chaotic mask and detect the original message. This synchronization is possible only when a similar chaotic circuit as that of sending end is fabricated. If the configuration circuit is secret, it is impossible to extract information from the transmitted message [5, 6]. Hence, there has been growing interest in the possibility of synchronizing chaotic signals. This idea has been tested theoretically as well as experimentally in the variety of dynamical system including Chua, Driven-Chua, and Chen circuits [7–9]. In general, the synchronization of dynamic chaotic oscillator has been realized via two main approaches; the design of feedback controllers to be tackled for the tracking problem related with the proper synchronization phenomena where oscillators with different order and structure can be synchronized and the design of state observers related with the synchronization of chaotic oscillators with equivalent order and topology [8]. The employment of chaotic signals in the wide field of secure communication has been important, because it was demonstrated that two chaotic systems starting from nonequal initial conditions can be synchronized if they are coupled [2, 6]. From the above, the use of observers which have been widely analyzed in control theory issues is one of the methodologies successfully employed for synchronization of

© Springer Nature Switzerland AG 2018
R. Martínez-Guerra and C. A. Pérez-Pinacho, *Advances in Synchronization of Coupled Fractional Order Systems*, Understanding Complex Systems,
https://doi.org/10.1007/978-3-319-93946-9_3

chaotic attractors of equivalent order and topology with applications to secure data transmission [5]. The main contribution of this chapter is to propose a master–slave synchronization scheme, where the systems to be synchronized have arbitrary initial conditions. In this configuration, the Chen oscillator is considered as an application example of a master system and the corresponding nonlinear observer is the slave system. The idea is that the trajectories of the observer follow the trajectories of the Chen oscillator. The observer structure contains a proportional and a bounded function of the synchronization error in order to provide asymptotic synchronization with a satisfactory performance. Numerical simulations are developed in order to provide the performance of the proposed synchronization methodology.

3.2   Application Case

The Chen system was proposed as an alternative chaotic attractor based on the structure of the Lorenz system, where the corresponding nonlinearities are related with a simple multiplicative terms of two state variables, however the Chen system depicted topological nonequivalent trajectories in comparison with the Lorenz oscillator [10]. The Chen oscillator has been applied to modeling simplified versions of brushless DC motors, lasers, thermosyphons, dynamos, chemical reactions, and electric circuits [11–14]. Chen dynamical system is described by a three order of autonomous ordinary differential equations as is shown in [15]:

$$\Sigma_m : \begin{cases} \dot{z}_1 = & \alpha(z - z_1) \\ \dot{z}_2 = (\gamma - \alpha) \; z_1 - \gamma z_2 - z_1 z_3 \\ \dot{z}_3 = \; z_1 \quad\quad z_2 - \beta z_3, \end{cases} \tag{3.1}$$

where the considered measured signal is $s = z_2$.

Here, z_1, z_2 and z_3 are the state variables and the parameters α, β, and γ are three positive real constants. This system contains a chaotic attractor when $\alpha = 35, \beta = 3$, and $\gamma = 28$. The trajectory of the system is specified by $(z_1(t), z_2(t), z_3(t))$. The critical points of the systems (3.1) are $CP_1 = (-\rho, -\rho, 2\gamma - \alpha); CP_2 = (\rho, \rho, 2\gamma - \alpha)$, and finally $CP_3 = (0, 0, 0)$, where $\rho = \sqrt{(\beta(2\gamma - \alpha))}$.

The divergence of the flow related with the system (3.1) is as follows:

$$\nabla \cdot F = \frac{\partial f_1}{z_1} + \frac{\partial f_2}{z_2} + \frac{\partial f_3}{z_3} = -\alpha + \gamma - \beta < 0,$$

when $\alpha + \beta > \gamma$.

Here, $F = (f_1, f_2, f_3) = (\alpha(z_2 - z_1), (\gamma - \alpha)z_1 - z_1 z_3 + \gamma z_2, z_1 z_2 - \beta z_3)$.

Thus, the system (3.1) is a forced dissipative system similar to a Lorenz system [16]. Thus, the solutions of the system (3.1) are bounded as $t \to \infty$. Chen shows that the system (3.1) exhibits chaos for specified values of the parameters $\alpha = 35$, $\beta = 3$, and $\gamma = 28$. Let us consider the system (3.1) as the master oscillator, in consequence the slave oscillator, under the master–slave synchronization scheme, is given by a Chen oscillator model disturbed by an external feedback, with the following structure:

$$u = k_0 e(t) + k_1 arctang(e(t))$$

From the above, the corresponding structure of the slave system is as follows:

$$f(\hat{z}) : \begin{cases} \dot{\hat{z}}_1 = \alpha(\hat{z}_2 - \hat{z}_1) + k_0 e(t) + k_1 arctan(e) \\ \dot{\hat{z}}_2 = (\gamma - \alpha)\hat{z}_1 - \gamma\hat{z}_2 - \hat{z}_1\hat{z}_3 + k_0 e(t) + k_1 arctan(e) \\ \dot{\hat{z}}_3 = \hat{z}_1\hat{z}_2 - \beta\hat{z}_3 + k_0 e(t) + k_1 arctan(e) \end{cases}$$

3.3 Synchronization Methodology

Now, considering the below general representation of a class of nonlinear system, as the chaotic oscillator

$$\dot{z} = f(z, u)$$
$$s = h(z) = Cz,$$

where $z \in \mathbb{R}^n$ is the variable states vector, $u \in \mathbb{R}^m$ is the control input $m \leq n$, and $s \in \mathbb{R}$ is the corresponding measured signal $f(z, u)$ is a differentiable vector function such that $f : \mathbb{R}^n \times \mathbb{R}^m \to \mathbb{R}^n$ it is Lipschitz continuous (with Lipschitz constant, $\Upsilon = \sup_{u \in \mathbb{R}^m, z \in \Omega \subset \mathbb{R}^n} |f'(z.u)| \geq 0$, being Ω a compact set), if and only if, it has bounded first derivative, one direction follows from the mean value theorem as follows:

$$\|f(z, u) - f(\hat{z}, u)\| \leq \Upsilon\|z - \hat{z}\| \tag{3.2}$$

Proposition 3.1 *The dynamic system acts as a slave system for the system* (3.1)

$$\dot{\hat{z}} = f(\hat{z}, u) + k_0 e(t) + k_1 arctan(e) \tag{3.3}$$

That is to say,

$$lim_{s \to \infty} |e(t)| = 0,$$

where $e = z - \hat{z}$ is defined as the synchronization error.

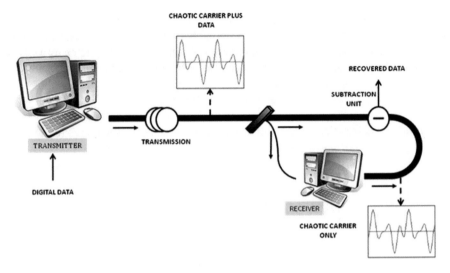

Fig. 3.1 Diagram of synchronization based secure data transmission

Note that the nonlinear feedback of the slave system satisfies the following property:

$$|arctan(e)| \leq \Gamma < \infty, \tag{3.4}$$

where Γ is a positive constant.

From the above, the dynamic modeling of the synchronization error is defined employing the Eqs. (3.1) and (3.3) as

$$\dot{e} = f(z, u) - f(\hat{z}, u) - k_0 e(t) - k_1 arctan(e) \tag{3.5}$$

In what follows an important observation is given

Remark 3.1 Let us define the output $s = z_2 + \omega$, where the output s is chaotic masking. The measurement is corrupted by a reasonable signal ω which is bounded, the information signal ω is embedded into the output of chaotic transmitter, the transmitter is a slight modification of the chaotic system (3.1). The schematic diagram of the chaotic communication based on slave system (receiver) is shown in Fig. 3.1. If the synchronization error is defined as $e = z - \hat{z}$, that is to say, in our case $e_2 = z_2 - \hat{z}_2$, where $\hat{z}_2$ represents the output of the receiver, then the recovered signal at receiver is given by $\hat{\omega} = s - \hat{s} = z_2 + \omega - \hat{z}_2 = e_2 + \omega$, with $\hat{s} = \hat{z}_2$. It should be noted that if e_2 tends to zero, the message (information signal) is completely recovered, that is to say, $\hat{\omega} = \omega$.

3.3.1 Convergence Analysis

Consider the Lyapunov candidate function

$$V = e^T Pe = ||e||_P^2, \ P = P^T \tag{3.6}$$

The time derivative along the trajectories of (3.6) is

$$
\begin{aligned}
\dot{V} &= \dot{e}^T Pe + e^T P\dot{e} \\
&= (f(z, u) - f(\hat{z}, u) - k_0 e(t) - k_1 arctan(e))^T Pe \\
&\quad + e^T P(f(z, u) - f(z, \hat{u}) - k_0 e(t) - k_1 arctan(e)) \\
&= 2e^T P(f(z, u) - f(\hat{z}, u)) - 2e^T P(k_0 e(t) + k_1 arctang(e))
\end{aligned}
\tag{3.7}
$$

(a) The matrix P can be expressed as $P = AA^T$, then

$$||e^T P(f(z, u) - f(\hat{z}, u))|| = ||e^T AA^T (f(z, u) - f(\hat{z}, u))|| = ||\tilde{e}^T \tilde{f}, || \tag{3.8}$$

where, $\tilde{e}^T = e^T A$ and $\tilde{f} = A^T (f(z, u) - f(\tilde{z}, u))$
Then,

$$
\begin{aligned}
||\tilde{e}^T || &= (\tilde{e}^T \tilde{e})^{1/2} \\
&= (e^T AA^T e)^{1/2} \\
&= (e^T Pe)^{1/2} = ||e||_P
\end{aligned}
\tag{3.9}
$$

As well, it is defined

$$||\tilde{f}|| = ||f||_P \tag{3.10}$$

Hence,

$$||e^T P(f(z, u) - f(\tilde{z}, u))|| = ||\tilde{e}^T \tilde{f}|| \leq ||\tilde{e}^T ||||\tilde{f}|| = ||e||_P ||f||_P \tag{3.11}$$

(b) Considering

$$||e^T P(k_0 e(t) + k_1 arctan(e)|| \leq ||k_0 e(t) + k_1 arctan(e)||||e||_P \tag{3.12}$$

From (a) and (b) and considering the bounded assumptions of the Chen oscillator and the nonlinear feedback, assumptions (3.2) and (3.4)

$$\dot{V} \le 2||e||_P \Upsilon ||e||_P - (k_1 \Gamma ||e||_P + k_0 ||e||_P^2) \qquad (3.13)$$

Then,

$$\dot{V} \le 2(\Upsilon - k_0)||e||_P^2 - k_1 \Gamma ||e||_P \qquad (3.14)$$

with an adequate choosing of the parameters k_0 and k_1, the following can be obtained $(\Upsilon - k_0) < 0$ and $k_1 > 0$.

$$\dot{V} \le 0$$

From the above, is concluded that the named synchronization error is stable. □.

3.4 Numerical Experiment and Results

Figure 3.1, shows a simplified diagram of a synchronization procedure under the master–slave structure. Numerical simulations were done in order to provide the performance of the proposed synchronization methodology; a personal computer (PC) with Intel Core $i7^T M$ processor and the ode solver from Matlab (*ode23s* library) were employed. For the master oscillator (3.1), the following initial condition was considered $z(0) = [1.5, 1.25, 7.5]$ and for the slave oscillator (3.3), the corresponding initial condition was $\hat{z}(0) = [1.2, 1, 5.8]$, the parameters of the master and the slave systems are the presented previously in Sect. 3.2. The synchronization procedure was turned on at time $t = 5$ s, the vector parameters of the corresponding feedback on the slave oscillator are $k_0 = [500, 500, 100]$ and $k_1 = [10, 10, 2]$ (Fig. 3.2). Figure 3.3 shows the variables of the observer z_1, z_2 and z_3 synchronized with the coordinates of the Chen oscillator z_1, z_2, and z_3 ($z = x$) respectively. In Fig. 3.4, it is illustrated as to how the attractors give us the synchronization or estimation errors, which tend to zero under the proposed methodology. Furthermore, a linear feedback observer is also implemented for comparison purposes, the linear observer contains the same vector of gains as the proposed methodology; as can be seen in Fig. 3.5, the corresponding synchronization errors are not enough diminished in comparison with the proposed nonlinear methodology, therefore, is concluded that the nonlinear synchronization procedure has a better performance. The proposed methodology is an adequate guess for real time implementation as shown in previous published works, where others nonlinear oscillators with similar structure were synchronized experimentally [7, 17].

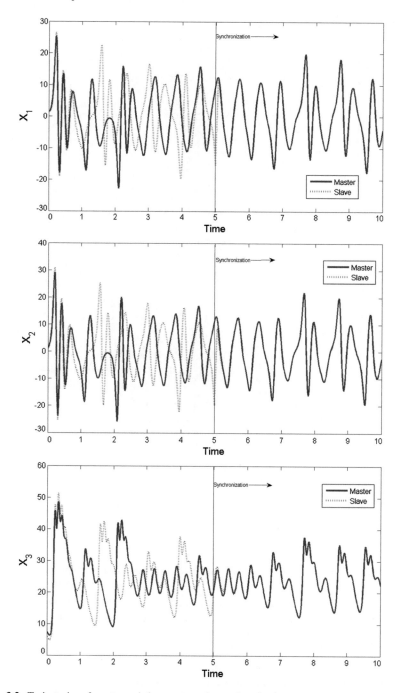

Fig. 3.2 Trajectories of master and slave systems in synchronization

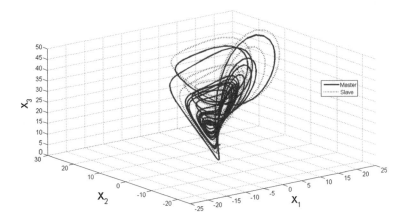

Fig. 3.3 Phase portrait of the master-slave synchronization

Fig. 3.4 Synchronization errors with the proposed observer

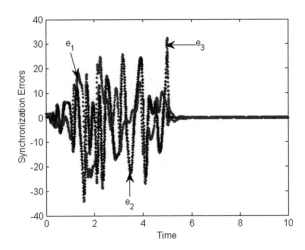

3.5 Conclusions

In this chapter, we tackled the master–slave synchronization problem via nonlinear observer design. We considered the Chen oscillator as an application case of the master and slave systems, the slave system corresponds with the proposed nonlinear observer, which contains a proportional and a bounded function of the synchronization error to guarantee asymptotic synchronization, the proposed methodology is applicable to a wide class of Lipschitz systems, where the corresponding nonlinear feedback of the slave system is bounded. Numerical experiments showed the performance of the proposed methodology and the linear (proportional) feedback observer.

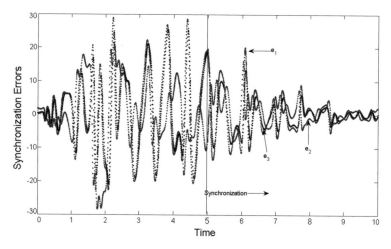

Fig. 3.5 Synchronization error with the linear feedback observer

References

1. Martínez-Guerra, Rafael, and Wen Yu, Chaotic synchronization and secure communication via sliding-mode observer. International Journal of Bifurcation and Chaos, 18(01), pp-235–243, (2008)
2. M.L. Hernault, J.P. Barbot, A. Ouslimani, IEEE Trans. Circ. Syst. I: Regular Papers 55, 614 (2008)
3. M. Mitra, S. Banerjee, Int. J. Mod. Phys. B 25, 521 (2011)
4. P. Saha, S. Banerjee, A. Roy Chowdhury, Phys. Lett. Sect. A: General, Atomic Solid State Phys. 326, 133 (2004)
5. Z.P. Jiang, IEEE Trans. Circ. Syst. I. Fund. Theory Appl. 44, 92 (2002)
6. K.M Cuomo, A.V. Oppenheim, Phys. Rev. Lett. 71, 65 (1993)
7. M.S. Baptista, S.P. Garcia, S.K. Dana, J. Kurths, Eur. Phys. J. Special Topics 165, 119 (2008).
8. M.T. Yassen, Chaos, Solitons Fractals 15, 271 (2003).
9. J.H. Lu, S.C. Zhang, Phys. Lett. A 286, 148 (2001)
10. G. Chen, T. Ueta, Int. J. Bif. Chaos 9, 1465 (1999)
11. N. Hemati, IEEE Trans. Circ. Syst. I: Fund. Theory Appl. 41, 40 (1994)
12. H. Haken, Phys. Lett. A 53,77 (1975)
13. E. Knobloch, Phys. Lett. A 82, 439 (1981)
14. D. Poland, Phys. D 65, 86 (1993)
15. J.H. Lu, T.S. Zhou, G. Chen, S.C. Zhang, Int. J. Bif. Chaos 12, 855 (2002a)
16. Martínez-Guerra, Rafael, G. C. Gómez-Cortés, and C. A. Pérez-Pinacho, Synchronization of Integral and Fractional Order Chaotic Systems, Springer, (2015).
17. R. Martínez-Guerra, D.M.G. Corona-Fortunio, J.L. Mata-Machuca, Appl. Math. Comput. 219, 10934 (2013)

Chapter 4
Synchronization for Chaotic System Through an Observer Using the Immersion and Invariance (I&I) Approach

In this chapter, the named master–slave configuration, where nonlinear systems observers and chaos synchronization are used together. The key idea is to design observers to accomplish chaos synchronization, where the slave is actually an observer coupled to the master through its corresponding output. This chapter aims at the master-slave synchronization by applying the Immersion and Invariance (I&I) method to solve the chaos synchronization problem for a kind of simple chaotic systems. To this end, a class of feedback-linearized chaotic systems is characterized. Afterwards, the I&I method is applied to propose the corresponding observer, or slave system, for such systems. This observer, which has some robust properties, allows asymptotic estimation of the underlying dynamics of the master system. Notably, the I& I approach has been successfully applied to control, identify, and observe a wide range of nonlinear systems. The seminal ideas of the I& I approach and its application can be found in [1, 2]. The chapter is organized as follows. In Sect. 4.1, the problem statement is established. In Sect. 4.2, the observer is proposed to solve the synchronization problem, by applying the I& I method. Section 4.3 shows the results of some numerical comparisons with other well-known observers. The conclusions are given in Sect. 4.4.

Notations and Definitions:

To simplify, we adopt the following notation

$$A_i = \begin{bmatrix} 0 & 1 & 0 & \dots & 0 \\ 0 & 0 & 1 & \dots & 0 \\ \vdots & \vdots & \vdots & & \vdots \\ 0 & 0 & 0 & \dots & 0 \end{bmatrix} \in \mathbb{R}^{i \times i}$$

© Springer Nature Switzerland AG 2018
R. Martínez-Guerra and C. A. Pérez-Pinacho, *Advances in Synchronization of Coupled Fractional Order Systems*, Understanding Complex Systems, https://doi.org/10.1007/978-3-319-93946-9_4

$$B_i = \begin{bmatrix} 0 \\ \dots \\ 1 \end{bmatrix} \in \mathbb{R}^i$$

$$C_i = \begin{bmatrix} 1 \\ \dots \\ 0 \end{bmatrix} \in \mathbb{R}^i$$

The vector $x_{k:n} = [x_k, \dots, x_n]^T \in \mathbb{R}^{n-k+1}$ stands for a sub-vector of vector $x = [x_1, x_2, \dots, x_n]^T \in \mathbb{R}^n$.

4.1 Preamble

Consider the following master chaotic configuration:

$$\Sigma_M : \begin{cases} \dot{x} = F(x, t) \\ y = h(x) \end{cases}, \tag{4.1}$$

where $x = [x_1, \dots, x_n]^T \in \mathbb{R}^n$ is the system state, $y \in \mathbb{R}$ is the single measurable output, and F and h are functions that depend on the arguments x_i. It is assumed that the vector fields F and h are forward complete [1]. That is, the trajectories starting at time t_0 are defined for all times $t > t_0$. The corresponding slave system Σ_S, with its state $\hat{x}(t, y) \in \mathbb{R}^{n_1}$, for the master system Σ_M, can be any kind of observer. Thus, the slave is considered synchronized with the master if

$$\lim_{t \to \infty} |x(t) - \hat{x}(t, y)| = 0$$

In this chapter, a new observer model is proposed, based on the I & I control methodology, to design the slave system for system (4.1). To this end, the class of chaotic system dealt with is characterized by introducing the following assumption (see [3]) :

Assumption A1: Suppose that there exists a primitive element y for system Σ_M that transforms it into the new form Σ_{M_T} given by

$$\Sigma_{M_T} : \begin{cases} \dot{z} = A_n z + B_n(\phi(y, t) + K^T z + f(z)) \\ y = C_n^T z \end{cases}, \tag{4.2}$$

where $z = [y, \dot{y}, \dots, y^{(n)}]^T \in \mathbb{R}^n$; with $z_k = y^{(k-1)}$, ϕ is a known scalar nonlinear function, which depends on y; K is a vector of constants, and $f(z)$ is a scalar Lipschitz function in z for the open set $D \subset \mathbb{R}^n$. That is,

$$||f(x) - f(y)|| \leq \gamma ||x - y||; \ \forall x, y \in D \tag{4.3}$$

Remark 4.1 The differential primitive element allows the formation of the observable canonical form (4.2) and several chaotic systems admit this representation. Examples of these systems include the Duffing chaotic oscillator, the Lorenz system, and the Genesio and Tesi system.

Problem Statement: Consider the system Σ_{M_T} defined in (4.2). The objective consists of designing and I & I-reduced observer for this system, such that

$$\lim_{t \to \infty} |z_{2:n}(t) - \tilde{z}(t, y)| = 0,$$

where $\tilde{z} \in \mathbb{R}^{n-1}$ is an estimation of the nonavailable state $z_{2:n} \in \mathbb{R}^{n-1}$, from the measurable output y.

Remark 4.2 Since y is available, it is not necessary to reconstruct the whole state $z = [y, z_{2:n}]^T$. This problem has been previously tackled and several solutions can be found in the literature, where many of the proposed observers therein (e.g., continuous observers, sliding observers, fuzzy observers, and high-gain observers) were designed into the basic idea of the Luenberger observer. Here, we propose a different solution, based on the Immersion and Invariance manifold approach, which provides a more general and abstract solution than the one based on the Luenberger Observer.

4.2 I&I Observer

In this section, the I&I observer methodology is adapted to recover asymptotically the underlying dynamics of system (4.2). To facilitate the design of this observer, the following additional assumption are made:
Assumption A2: There exists a mapping $r(y) \in \mathbb{R}^{n-1}$, defined as

$$r(y) = y r_d + \int_0^y \rho(y(s)) ds, \tag{4.4}$$

where $r_d = [r_{d_2}, r_{d_3}, \dots, r_{d_n}]^T$ is a constant vector, and $\rho(y) = [\rho_2(y), \dots, \rho_n(y)]^T$ is a vector of variable functions.

Assumption A3: Let $M(y) = M_K + M_1(y) \in \mathbb{R}^{(n-1) \times (n-1)}$, where M_K is a constant matrix, and $M_1(y)$ is a matrix of variable functions, defined as

$$M_K = -r_d C_{(n-1)}^T + A_{(n-1)} + B_{(n-1)} K_{2:n}^T + M_1(y) \tag{4.5}$$

$$M_1(y) = -\rho(y) C_{(n-1)}^T$$

Then, there exist $P_{(n-1)} = P_{(n-1)}^T > 0$, such that

$$P_{(n-1)} M_K + M_K^T P_{(n-1)} = -I_{(n-1)} \tag{4.6}$$

and

$$\Psi(y) = M_1^T(y) P_{(n-1)} + P_{(n-1)} M_1(y) \leq 0 \tag{4.7}$$

That is, M_K is selected as a Hurwitz matrix, and $\Psi(y) \leq 0$.

Now, the main proposition of this work can be presented as follows:

Proposition 4.1 *Considers systems (4.2) under assumptions A2 and A3. Then, the following I& I observer*

$$\tilde{z} = \hat{q} + r(y) \in \mathbb{R}^{n-1}$$

assures global exponential convergence of the observation error, if $2\gamma \|\bar{r}\| < 1$, with $\bar{r} = |P_{(n-1)} B_{(n-1)}|$, and $r(y)$ is obtained from (4.4), and the dynamic of $\hat{q}$ is computed via (4.15). That is, for all initial conditions $z(0)$, there exist α and $\lambda \in \mathbb{R}^+$, such that

$$|\tilde{z}(t) - z_{2:n}(t)| \leq \alpha e^{-\lambda t} \tag{4.8}$$

Proof To this end, the I& I procedure given in [1] is followed verbatim. Let us define the $(n-1)$ dimensional vector error as

$$e = \tilde{z} - z_{2:n} = \hat{q} + r(y) - z_{2:n}, \tag{4.9}$$

where the dynamic of vector $\hat{q}$ is proposed below in expression (4.11). The objective of the I & I method is to select $\hat{q}$ and the mapping $r(y)$ to assure an asymptotic stable dynamic for the observation error e. Differentiating e, defined in (4.9), and noting that $\dot{y} = C_{(n-1)}^T z_{2:n}$, the following is obtained

$$\dot{e} = \dot{\hat{q}} + C_{(n-1)}^T z_{2:n} \frac{dr}{dy} - \dot{z}_{2:n} \tag{4.10}$$

Evidently, form the first equation of (4.2), we have that $\dot{z}_{2:n}$ can be expressed as

$$\dot{z}_{2:n} = A_{(n-1)} z_{2:n} + B_{(n-1)} (\phi(y, t) + f(y, z_{2:n}) + k_1 y + KZ) \tag{4.11}$$

$$KZ = K_{2:n}^T z_{2:n}$$

Therefore, after substituting the values of (4.11) into Eq. (4.10) and using assumptions **A2** and **A3**, we obtain the following equation

$$\dot{e} = \dot{\hat{q}} - M(y)z_{2:n} - B_{(n-1)}(f(y, z_{2:n}) - f(y, \tilde{z})) - B_{(n-1)}(\phi(y, t) + k_1 y + f(y, \tilde{z})),$$
(4.12)

where $M(y)$ is given by

$$M(y) = -(r_d + \rho(y))C_{(n-1)}^T + A_{(n-1)} + B_{(n-1)}K_{2:n}^T$$
(4.13)

Notice that the definition of $M(y) = M_K + M_1(y)$, given in A3, is in agreement with Eq. (4.13). From Eq. (4.9), $z_{2:n} = \tilde{z} - e$. Now, substituting these values into Eq. (4.12), leads to expression:

$$\dot{e} = \dot{\hat{q}} + M(y)e - B_{(n-1)}(f(y, \tilde{z} - e) - f(y, \tilde{z})) - M(y)\tilde{z} - B_{(n-1)}(\phi(y, t) + k_1 y + f(y, \tilde{z}))$$
(4.14)

Therefore, $\dot{\hat{q}}$ can be fixed as

$$\dot{\hat{q}} = M(y)\tilde{z} + B_{(n-1)}(\phi(y, t) + k_1 y + f(y, \tilde{z}))$$
(4.15)

and Eq. (4.14) becomes

$$\dot{e} = M(y)e - B_{(n-1)}(f(y, \tilde{z} - e) - f(y, \tilde{z}))$$
(4.16)

To analyze the convergence of vector e, the following Lyapunov function is used

$$V = e^T P_{(n-1)}e$$

According to Eq. (4.6), the time derivative of V along of the trajectory of system (4.16) leads to

$$\dot{V} = -|e|^2 + 2e^T P_{(n-1)}B_{(n-1)}(f(y, \tilde{z}_{2:n} - e) - f(y, \tilde{z}_{2:n})) + E, \quad E = e^T \Psi(y)e$$
(4.17)

From inequality (4.7), it is claimed here that $\Psi(y) \leq 0$. On the other hand, since f is a scalar Lipschitz function, it follows that $\dot{V}$ can be upper bounded by

$$\dot{V} \leq -|e|^2 + 2\gamma \bar{r}|e|^2,$$
(4.18)

where $\bar{r} = |P_{(n-1)}B_{(n-1)}|$. Therefore, selecting $2\gamma||\bar{r}|| < 1$, it is assured that e asymptotically and exponentially converges to zero. Finally, remembering that $e = \tilde{z} - z_{2:n}$ (see (4.9)), it can be claimed that $\tilde{z}$ is an asymptotic estimator for the vector state $z_{2:n}$.

Remark 4.3 From the definition of (4.5), matrices M_K and $M(y)$ are explicitly computed as

$$M_K = \begin{bmatrix} -rd_2 & 1 & 0 & \dots & 0 \\ -rd_3 & 0 & 1 & \dots & 0 \\ \vdots & \vdots & \vdots & & \vdots \\ k_2 - rd_n & k_3 & \dots & k_n, & \end{bmatrix} \tag{4.19}$$

$$M_1(y) = \begin{bmatrix} -p_2(y) & 0 & 0 & \dots & 0 \\ -p_3(y) & 0 & 0 & \dots & 0 \\ \vdots & \vdots & \vdots & & \vdots \\ -p_n(y) & 0 & \dots & 0, & \end{bmatrix} \tag{4.20}$$

where the characteristic polynomial of M_K is given by

$$P_K(s) = s^{n-1} + s^{n-2}(rd_2 - k_n) + PP$$
$$PP = s^{n-3}(rd_3 - k_n rd_2 - k_{n-1}) + \cdots + PPP$$
$$PPP = (rd_n - k_n rdn - 1 - k_{n-1} rd_{n-2} - \cdots - k_2)$$

implying that $P_K(s)$ can be Hurwitz, if constants $rd_i > 0$, with $i \in l_{2:n}$, are conveniently fixed. On the other hand, if vector $K = 0$, it is easily concluded that the characteristic polynomial of matrix $M_{K=0} = M_0$ is given by

$$P_0(s) = s^{n-1} + s^{n-2}rd_2 + s^{n-3}rd_3 + \cdots + rd_n \tag{4.21}$$

This section ends with analyzing the robust property of the proposed approach. To this end, the following assumption is made.

Assumption A4 The trajectories of systems (4.2) belong to a bounded set $D \in \mathbb{R}^n$, and there exists $\delta > 0$, such that

$$\sup_{z \in D} \left| \underbrace{\phi(y, t) + K^T z + f(z)}_{W(z)} \right| \le \delta \tag{4.22}$$

Following the steps in Proposition 4.1, Eq. (4.14), evidently, becomes:

$$\dot{e} = \dot{\hat{q}} + (M_0 + M_1(y))e - B_{(n-1)}W(z) - (M_0 + M_1(y))\tilde{z} \tag{4.23}$$

with M_0 and $M_1(y)$ defined previously in Remark 4.3. Therefore, fixing the dynamic of $\hat{q}$ as

$$\dot{\hat{q}} = (M_0 + M_1(y))\tilde{z} \tag{4.24}$$

the error dynamic is given by

$$\dot{e} = (M_0 + M_1(y))e - B_{(n-1)}W(z) \tag{4.25}$$

Now, M_0 and $M_1(y)$ are selected, such that Eqs. (4.6) and (4.7) are fulfilled, then the time derivative of $V = e^T P_{(n-1)}e$ leads to the following equation.

$$\dot{V} = -|e|^2 + 2e^T P_{(n-1)} B_{(n-1)} W(z) + e^T \Psi(y)e \tag{4.26}$$

that can be upper bounded, as

$$\dot{V} = -|e|^2 + 2\bar{r}\delta|e|$$

From the last equation, it follows that $|e(t)| \leq 1/(2\bar{r}\delta)$ for all $t > t_* > 0$, where t_* is a finite time. That is, the error is ultimately bounded stable.

Notice that constant r can be as small as desired, if the roots of the characteristic polynomial $P_0(s)$, given in (4.21), are adequately fixed. For instance, in a first step, the Immersion & Invariance Observer (I&IO) can be made to behave as a reduced high-gain observer. It can be accomplished by fixing $M_{0\mu}$ and $M_{1\mu}(y)$ as

$$M_{0\mu} = \begin{pmatrix} -\dfrac{d_2}{\mu} & 1 & 0 & \dots & 0 \\ -\dfrac{d_3}{\mu^2} & 0 & 1 & \dots & 0 \\ \vdots & \vdots & \vdots & \vdots & \vdots \\ -\dfrac{d_n}{\mu^n} & 0 & 0 & \dots & 0, \end{pmatrix} \tag{4.27}$$

$$M_{1\mu}(y) = \begin{bmatrix} -\dfrac{\rho_2(y)}{\mu} & 0 & \dots & 0 \\ -\dfrac{\rho_3(y)}{\mu^2} & 0 & \dots & 0 \\ \vdots & \vdots & \vdots & \vdots \\ -\dfrac{\rho_n(y)}{\mu^n} & 0 & \dots & 0, \end{bmatrix} \tag{4.28}$$

where $0 < \mu < 1$ and $p(s) = s^{n-1} + d_2 s^{n-2} + \dots + d^n$ is Hurwitz, and both matrices are restricted to

$$P_{(n-1)}M_{01} + M_{01}^T P_{(n-1)} M_{11}(y) \leq 0, \quad M_{11}(y) = M_{1(\mu=1)}(y) \tag{4.29}$$

and

$$\Psi(y) = M_{11}^T(y)P_{(n-1)} + P_{(n-1)}M_{11}(y) \leq 0 \qquad (4.30)$$

with $P_{(n-1)} > 0$. Second, if we redefine the error as $e' = D(\mu)e$ (where $D(\mu)$ is a diagonal matrix that can be found in several high-gain observer schemes) it is easy to see the following the steps in [4], that

$$|e'(t)| \leq k_\delta \mu \quad \text{for} \quad t \geq t_* > 0,$$

where k_δ is a constant that depends on both δ and d_i. That is, e' is bounded by the constant $k_\delta\mu$, where μ can be as small as desired. The last discussion and developments are summarized in the following proposition.

Proposition 4.2 *Consider system (4.2), under assumption A2 and A4. If matrices $M_{0\mu}$ and $M_{1\mu}$ are selected according to (4.29) and (4.30), respectively, then the following I&I reduces high-gain observer (I&IHGO)*

$$\tilde{z} = \hat{q} + r(y)$$

with $r(y)$ and $\hat{q}$ given, respectively, in (4.4) and (4.24), assures that the error is ultimately bounded and stable.

4.3 Numerical Evaluations

In this section, the effectiveness of the proposed I&IO approach is illustrated when applied to solving the chaotic synchronization problem for the class of chaotic systems defined in (4.2). To this end, numerical simulations are used to compare the proposed approach to the performances of a well established high-gain observer and the Luenberger observer, the latter being a particular case of the former. In the experiment, the Duffings oscillator is used. Finally, we mention that a formal comparison between these observers performances is beyond the scope of this work, because optimal control arguments should be considered.

4.3.1 Duffing's Mechanical Oscillator

This system is an example of a periodically forced oscillator with a nonlinear elasticity, describe by the following equations [5]:

$$\dot{z}_1 = z_2$$
$$\dot{z}_2 = -p_1 z_2 - p_3 z_1^3 + p_2 z_1 + A\cos(\omega t),$$

where measurable position is defined as $y = z_1$ and z_2 is the nonavailable system velocity. It is well known that in a neighborhood of $\{A = 0.3, \ p_1 = 0.2, \ p_2 = -1.1, \ p_3 = 1, \omega = 1\}$, this system exhibits a chaotic behavior. Naturally, the Duffing system belongs to the class of chaotic systems defined in (4.2), where

$$K = [0, -p_1]^T; \quad \phi(y, t) = -p_3 y^3 + p_2 y + A\cos(\omega t); \quad f(z) = 0.$$

Using Remark 4.3, it follows that $M_k = -p_1 - r_2$ and $M_1(y) = -\rho_2(y)$, where r_2 and ρ are control parameters selected according to Eqs. (4.6) and (4.7). Notice that assumption A3 is assured if $P_1 = 1, r_2 = -p_1 + 1/\epsilon > 0$, with $0 < \epsilon < 1$ and $\rho_2(y) = \rho(y)$, where $\rho(y)$ is any positive function. For this particular case, $\rho(y) = \bar{k}_1 sech(y) + \bar{k}_2 y^{2i}; \ \bar{k}_1, \bar{k}_2 \geq 0$ and $i \in \mathbb{N}$ is selected. Consequently, the corresponding I &IO is given by

$$\bar{z} = \hat{q} + \frac{1}{\epsilon} y + \bar{k}_1 \tan^{-1}(tanh(\frac{y}{2})) + \bar{k}_2 y^{2i+1},$$

where $\hat{q}$ is generated by

$$\dot{\hat{q}} = -\left(\frac{1}{\epsilon} + \bar{k}_1 sech(y) + \bar{k}_2 y^{2k}\right) \bar{z} + \phi(y, t)$$

Numerical experiment set-up: The Duffing's system parameters were chosen to be the same as in the nominal case. The initial conditions were fixed as $z_1(0) = 1$ and $z_2(0) = -1$. The corresponding values of the I&IO were $\bar{k}_1 = 0.5, \bar{k}_2 = 0.1$, and $\epsilon = 0.5$. The high-gain parameters were selected such that $p(s) = (s + 1)^2$ and $\mu = 0.2$. Finally, the Luenberger observer parameters were selected, such that its characteristic polynomial was fixed to be $p(s) = (s + 1)^2$. To compare the effectiveness of these observers a numerical test was carried out, where the parameters are known with 90%

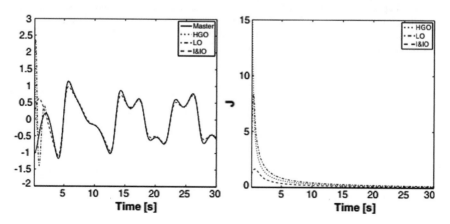

Fig. 4.1 Numerical comparison, parameters are known with 90% accuracy

accuracy. The results are shown in Fig. 4.1. For this set-up, the proposed approach slightly outperforms the other observers, with the exception of high-gain observer which has better performance.

4.4   Conclusions

This work presents a novel solution to the synchronization problem in the master–slave configuration for a class of feedback-linearized chaotic systems. This class of systems have the advantage of having a structure similar to a cascade chain of integrators. This structure allows application of the I&I method to design the corresponding observer or slave system. The obtained observer exhibits some robust properties and can be considered as a generalized observer, because it can behave as a reduced-order, high-gain observer, under some considerations. To establish the efficiency of the I&IO, some numerical experiments were designed in order to compare this observer and the high-gain and Luenberger observer. In the obtained results, the proposed observer slightly outperforms the other two.

References

1. Alessandro Astolfi and Romeo Ortega. Immersion and invariance: a new tool for stabilization and adaptive control of nonlinear system. Automatic Control, IEEE Transactions on, 48(4):590–606, 2003.
2. Xiangbing Liu, Romeo Ortega, Hongye Su, and Jian Chu. Immersion and invariance adaptive control of nonlinearly parameterized nonlinear systems. Automatic Control, IEEE Transaction on 55(9):2209–2214, 2010.
3. Martínez-Guerra, Rafael, and Christopher Diego Cruz-Ancona. Algorithms of Estimation for Nonlinear Systems: A Differential and Algebraic Viewpoint. Springer, 2017.
4. AN Atassi and HK Khalil. Separation results for the stabilization of nonlinear systems using different high-gain observer designs. Systems & Control Letters, 39(3): 183–191, 2000.
5. J Heagy and WL Ditto. Dynamics of a two-frequency parametrically driven duffing oscillator. Journal of Nonlinear Science, 1(4):423–455, 1991.

Chapter 5
Synchronization of Nonlinear Fractional-Order Systems by Means of $PI^{r\alpha}$ Reduced Order Observer

5.1 Introduction

The main contribution in this chapter is the synthesis of a new fractional-reduced-order observer for the synchronization problem in partially known nonlinear fractional-order systems, we propose a $PI^{r\alpha}$ reduced-order observer for estimating the unknown state variables based on Fractional Algebraic Observability (FAO) property (a system's copy is not necessary). This novel observer presents some advantages, for example, the norm of the estimation error, the time of convergence, and the performance of the $PI^{r\alpha}$ reduced-order observer can be improved by the correct choice of the gains.

5.2 Problem Statement and Main Result

We take the initial condition problem for an autonomous fractional order nonlinear system, with $0 < \alpha < 1$:

$$x^{(\alpha)} = f(x), \quad D^{\alpha-1}x(0) = (I_{0+}^{1-\alpha}x)(0+) = x_0 \qquad (5.1)$$
$$y = h(x),$$

where $x \in \Omega \subset \mathbb{R}^n$, $f : \Omega \to \mathbb{R}^n$ is a Lipschitz continuous function (this assures the unique solution [1]), with $x_0 \in \Omega \subset \mathbb{R}^n$, in this case, y denotes the output of the system (the measure that we can obtain), $h : \mathbb{R}^n \to \mathbb{R}^q$ is a continuous function and $1 \leq q \leq n$.

Consider the system given by (5.1), we will separate in two dynamical systems with states $\bar{x} \in \mathbb{R}^p$, which represents the states that we can obtain directly

© Springer Nature Switzerland AG 2018
R. Martínez-Guerra and C. A. Pérez-Pinacho, *Advances in Synchronization of Coupled Fractional Order Systems*, Understanding Complex Systems, https://doi.org/10.1007/978-3-319-93946-9_5

via algebraic relations of the output (known states), and $\eta \in \mathbb{R}^{n-p}$, respectively with $x^T = (\bar{x}^T, \eta^T)$, the first system will describe the known states and the second represents unknown states, then the system (5.1) can be written as

$$\bar{x}^{(\alpha)} = \bar{f}(\bar{x}, \eta),$$
$$\eta^{(\alpha)} = \Delta(\bar{x}, \eta) \tag{5.2}$$
$$y = h(\bar{x}),$$

where $f^T(x) = (\bar{f}^T(\bar{x}, \eta), \Delta^T(\bar{x}, \eta))$, $\bar{f} \in \mathbb{R}^p$, $h : \mathbb{R}^p \to \mathbb{R}^q$ and $\Delta \in \mathbb{R}^{n-p}$ with $1 \leq p \leq q \leq n$. Then, we only need to estimate the η's states.

The components of unknown state vector η are assumed to be FAO (see Definition 2.4), and the problem in terms of the master-slave synchronization scheme is defined in the following manner.

Let us consider the master system:

$$\eta_i^{(\alpha)} = \Delta_i(x, \eta), \tag{5.3}$$

$$y_{\eta_i} = \eta_i = \phi_i \left(y_{\bar{x}}, y_{\bar{x}}^{(\alpha)}, D^{(2\alpha)} y_{\bar{x}}, \cdots, D^{(r\alpha)} y_{\bar{x}} \right) \tag{5.4}$$

for $p + 1 \leq i \leq n$, where η_i is a component of the state vector η and y_{η_i} denotes the output of the ith master system.

The proposed observer (slave system) is given by

$$\hat{\eta}_i^{(\alpha)} = K_{i0}(\eta_i - \hat{\eta}_i) + \sum_{j=1}^{r'} K_{ij}(\eta_i^{(-j\alpha)} - \hat{\eta}_i^{(-j\alpha)}), \quad D^{\alpha-1}\hat{\eta}_i(0) = \hat{\eta}_{i0} \in \mathbb{R} \tag{5.5}$$

$$y_{\hat{\eta}_i} = \hat{\eta}_i, \tag{5.6}$$

for $p + 1 \leq i \leq n$, where $\hat{\eta}_i$ is the state, $y_{\hat{\eta}_i}$ denotes the output of the i-th slave system and the gains K_{ij}, $j = 0, 1, \ldots, r'$ will be selected in order to fulfill the stability conditions for the observer.

Given the master system (5.3) and the slave system (5.5), it should be determined some conditions, such that the output of the slave system (5.6) synchronizes with the output of the master system (5.4).

The synchronization error can be defined as

$$e_i = y_{\eta_i} - y_{\hat{\eta}_i} = \eta_i - \hat{\eta}_i \tag{5.7}$$

Let us assume the following conditions:
H1. η_i satisfies the FAO property for $p + 1 \leq i \leq n$.
H2. Δ_i is bounded, i.e., $\exists N \in \mathbb{R}^+$ such that $\|\Delta_i(x)\| \leq N$, $N < \infty$, $\forall x \in \Omega$.
H3. The gains K_{ij}, $j = 0, 1, \ldots, r'$, are chosen such that $P_i(s^\alpha)$ is stable.

Now, we are in position to establish the following result related with the convergence analysis of the error.

Theorem 5.1 *Let the system (5.1) which can be expressed as (5.2), if the conditions H1-H3 are satisfied then the synchronization of the master output (5.4) with the slave output (5.6) is achieved, for global initial conditions of the states.*

Proof The proposed observer is given by

$$\hat{\eta}_i^{(\alpha)} = K_{i0}(\eta_i - \hat{\eta}_i) + \sum_{j=1}^{r'} K_{ij}(\eta_i^{(-j\alpha)} - \hat{\eta}_i^{(-j\alpha)}) \tag{5.8}$$

Let the error be $e_i = \eta_i - \hat{\eta}_i$ then,

$$e_i^{(\alpha)} = \eta_i^{(\alpha)} - \hat{\eta}_i^{(\alpha)} \tag{5.9}$$

Substituting (5.8) in (5.9), the following equation for the error is obtained:

$$e_i^{(\alpha)} = \eta_i^{(\alpha)} - K_{i0}(\eta_i - \hat{\eta}_i) - \sum_{j=1}^{r'} K_{ij}(\eta_i^{(-j\alpha)} - \hat{\eta}_i^{(-j\alpha)}) \tag{5.10}$$

Taking the Laplace transform to both sides of Eq. (5.10)

$$s^\alpha E_i(s) - [D^{\alpha-1}e_i(0)] = s^\alpha H_i(s) - [D^{\alpha-1}\eta_i(0)] \tag{5.11}$$

$$- K_{i0}E_i(s) - \sum_{j=1}^{r'} K_{ij}s^{-j\alpha}E_i(s)$$

Then,

$$E_i(s) = \frac{\Psi_i(s^\alpha)}{P_i(s^\alpha)},$$

where
$\Psi_i(s^\alpha) = s^{r\alpha}[s^\alpha H_i(s) - (D^{\alpha-1}\eta_i(0)) + D^{\alpha-1}e_i(0)]$ and $P_i(s^\alpha) = s^{(r+1)\alpha} + K_{i0}s^{r\alpha} + K_{i1}s^{(r-1)\alpha} + ... + K_{i(r'-1)}s^\alpha + K_{ir'}$.

Note that it is possible to obtain a factorization of $P_i(s^\alpha)$ as

$$P_i(s^\alpha) = \prod_{j=1}^{r+1}(s^\alpha + \lambda_j)$$

Furthermore, the condition of stability for the polynomial $P_i(s^\alpha)$ is given by

$$|arg(\lambda_j)| > \alpha \frac{\pi}{2}$$

where λ_j is the jth root of the polynomial $P_i(s^\alpha)$ (see [2]).

Hence, the norm of the error is given as follows:

$$\| E_i(s) \| = \| \frac{\Psi_i(s^\alpha)}{P_i(s^\alpha)} \| = \| \sum_{j=1}^{r+1} \frac{l_{ij}}{(s^\alpha + \lambda_j)} \| \le \sum_{j=1}^{r+1} \| \frac{l_{ij}}{(s^\alpha + \lambda_j)} \|, \quad l_{ij} \in \mathbb{R} \quad (5.12)$$

Thus, the error is bounded.

Finally, if the values of K_{ij}, $j = 0, 1, \ldots, r'$ are chosen, such that, $P_i(s^\alpha)$ is stable (H3), then the norm of the error can be arbitrarily reduced with an appropriate selection of K_{ij}.

Remark 5.1 If the FAO of state variable is expressed in terms of fractional sequential derivatives of the output y, which are unknown, then it is necessary to introduce an artificial variable (if it is possible) in order to avoid the use of these unknown derivatives.

Remark 5.2 Chaotic systems are characterized by global boundedness of the trajectories [3], then H2 is always satisfied.

5.3   Numerical Results

In this section, the synchronization of nonlinear fractional-order systems is studied by means of numerical simulations. The fractional-order Rössler hyperchaotic system is presented as an example.

First, consider fractional-order Rössler hyperchaotic system [4].

$$x^{(\alpha)} = \begin{bmatrix} x_3 + ax_1 + x_2 \\ -cx_4 + dx_2 \\ -x_1 - x_4 \\ b + x_3x_4, \end{bmatrix} \quad (5.13)$$

where $x = (x_1, x_2, x_3, x_4)^T$ is the state vector, y_1 and y_2 are the outputs. When $a = 0.38, b = 3, \alpha = 0.95$, the Rössler equation (5.13) has a hyperchaotic attractor.

Now, the system (5.13) could be rewritten in the form (5.2) as follows:

$$\bar{x}^{(\alpha)} = \begin{bmatrix} \eta_3 + a\bar{x}_1 + \bar{x}_2 \\ -c\eta_4 + d\bar{x}_2, \end{bmatrix}$$

$$\bar{\eta}^{(\alpha)} = \begin{bmatrix} -\bar{x}_1 - \eta_4 \\ b + \eta_3\eta_4, \end{bmatrix} \tag{5.14}$$

$$y = \begin{bmatrix} \bar{x}_1 \\ \bar{x}_2, \end{bmatrix}$$

where $y_1 = x_1 = \bar{x}_1$, $y_2 = x_2 = \bar{x}_2$, $\eta_3 = x_3$, $\eta_4 = x_4$. From (5.14), it is possible to find the following relations:

$$\eta_3 = \phi_3(y_{\bar{x}}, y_{\bar{x}}^{(\alpha)}) = y_1^{(\alpha)} - ay_1 - y_2 \tag{5.15}$$

$$\eta_4 = \phi_4(y_{\bar{x}}, y_{\bar{x}}^{(\alpha)}) = -\frac{1}{c}y_2^{(\alpha)} + \frac{d}{c}y_2 \tag{5.16}$$

Then, $\eta_3 = x_3$ and $\eta_4 = x_4$ are FAO and therefore H1 is fulfilled. From above, the master system are given by

$$\eta_3^{(\alpha)} = -\bar{x}_1 - \eta_4$$
$$\eta_3 = y_1^{(\alpha)} - ay_1 - y_2 \tag{5.17}$$

$$\eta_4^{(\alpha)} = b + \eta_3\eta_4$$
$$\eta_4 = -\frac{1}{c}y_2^{(\alpha)} + \frac{d}{c}y_2 \tag{5.18}$$

For this example $r = 1$, i.e., we use a PI^α reduced order observer.

Now we design the corresponding slave system for (5.17) and (5.18) for master system (5.17) and using (5.5) we have

$$\hat{\eta}_3^{(\alpha)} = K_{3,0}(\eta_3 - \hat{\eta}_3) + K_{3,1}D^{-\alpha}(\eta_3 - \hat{\eta}_3) \tag{5.19}$$

$K_{3,j}$, $j = 0, 1$ are selected such that $P(s^\alpha) = s^{2\alpha} + K_{3,0}s^\alpha + K_{3,1}$ be stable.

Using (5.17) and after some algebraic manipulations we have

$$\gamma_3^{(\alpha)} = \begin{aligned} & K_{3,0}(-ay_1 - y_2 - (\gamma_3 + K_{3,0}y_1)) \\ & +K_{3,1}y_1 + K_{3,1}D^{-\alpha}(-ay_1 - y_2 - (\gamma_3 + K_{3,0}y_1), \end{aligned} \tag{5.20}$$

where $\gamma_3 = \hat{\eta}_3 - K_{3,0} y_1$ is an auxiliary variable introduced in order to avoid the use of derivatives of y_1. Now consider Eq. (5.18) and by an analogous procedure we have:

$$\hat{\eta}_4^{(\alpha)} = K_{4,0}(\eta_4 - \hat{\eta}_4) + K_{4,1} D^{-\alpha}(\eta_4 - \hat{\eta}_4), \tag{5.21}$$

where $K_{4,j}$, $j = 0, 1, 2$ are selected such that $P(s^\alpha) = s^{3\alpha} + K_{4,0} s^{2\alpha} + K_{4,1} s^\alpha + K_{4,2}$ be stable.

$$\begin{aligned}
\gamma_4^{(\alpha)} = &\ K_{4,1}\left(\frac{d}{c} y_2 - (\gamma_4 - \frac{K_{4,1}}{c} y_1)\right) - \frac{K_{4,2}}{c} y_2 \\
&+ K_{4,2} D^{-\alpha}\left(\frac{d}{c} y_2 - (\gamma_4 - \frac{K_{4,1}}{c} y_1)\right),
\end{aligned} \tag{5.22}$$

where $\gamma_4 = \hat{\eta}_4 + \dfrac{K_{4,1}}{c} y_1$. Then, it is possible to obtain the estimates $\hat{\eta}_3$ and $\hat{\eta}_4$ from the following relations:

$$\begin{aligned}
\hat{\eta}_3 &= \gamma_3 + K_{3,0} y_1 \\
\hat{\eta}_4 &= \gamma_4 - \frac{K_{4,1}}{c} y_1
\end{aligned} \tag{5.23}$$

Numerical simulations were performed with the following parameters: $a = 0.38$, $b = 3$, $c = 0.5$, $d = 0.05$, $\alpha = 0.95$ with initial conditions for the observer (slave system) given by $\hat{\eta}_{30} = -1800$ and $\hat{\eta}_{40} = -9000$ and initial conditions for the master system $x_0 = [-30, 60, -20, 20]^T$. The convergence of the estimates to the true signals is shown in Fig. 5.1.

Now, consider fractional-order Lorenz system

$$x^{(\alpha)} = \begin{bmatrix} ax_2 - ax_1 \\ bx_1 - cx_2 - x_1 x_2 \\ x_1 x_2 - dx_3 \end{bmatrix}, \tag{5.24}$$

where $x = [x_1, x_2, x_3]^T$ is the state and the output is y_1. When $a = 10$, $b = 28$, $c = -2$, $d = 8/3$, $\alpha = 0.96$, the Lorenz system (5.24) exhibits a chaotic behavior.

It is possible to rewrite system (5.24) in the form (5.2) as follows:

$$\bar{x}^{(\alpha)} = a\eta_2 - a\bar{x}_1,$$

$$\eta^{(\alpha)} = \begin{bmatrix} b\bar{x}_1 - c\eta_2 - \bar{x}_1 \eta_3 \\ \bar{x}_1 \eta_2 - d\eta_3 \end{bmatrix}, \tag{5.25}$$

$$y = \bar{x}_1,$$

where $y_1 = x_1 = \bar{x}_1$, $\eta_2 = x_2$, $\eta_3 = x_3$. From (5.25), the following relations are achieved:

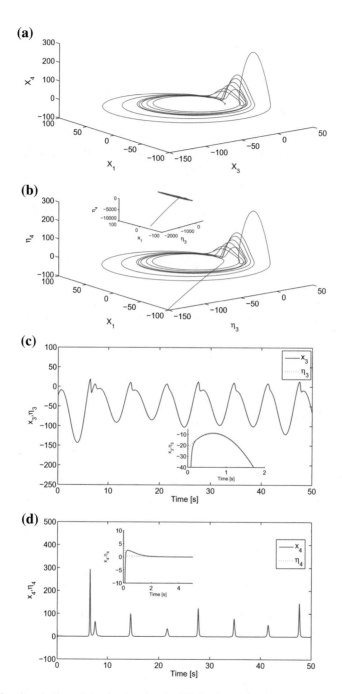

Fig. 5.1 Synchronization of the fractional-order Rössler hyperchaotic system, **a** $x_3 - x_1 - x_4$ space **b** $\hat{\eta}_3 - x_1 - \hat{\eta}_4$ space **c** and **d** shows the convergence of the estimates $\hat{\eta}_3$ and $\hat{\eta}_4$ to the states x_3 and x_4

$$\eta_2 = \phi_2(y_{\bar{x}}, y_{\bar{x}}^{(\alpha)}) = \frac{1}{a} y_1^{(\alpha)} + y_1 \tag{5.26}$$

$$\eta_3 = \phi_3(y_{\bar{x}}, y_{\bar{x}}^{(\alpha)}, D^{(2\alpha)} y_{\bar{x}}) = -\frac{1}{a y_{\bar{x}}} [D^{(2\alpha)} y_{\bar{x}} + (a+c) y_{\bar{x}}^{(\alpha)} + a(c-b) y_{\bar{x}}] \tag{5.27}$$

Note that η_3 looses algebraic observability property when $y_{\bar{x}} = x_1 = 0$, hence, only (5.26) satisfies FAO with respect to the selected output $y_{\bar{x}} = x_1$.

Then, from (5.26), we obtain the following master system:

$$\eta_2^{(\alpha)} = b\bar{x}_1 - c\eta_2 - \bar{x}_1 \eta_3$$

$$\eta_2 = \frac{1}{a} y_1^{(\alpha)} + y_1 \tag{5.28}$$

Now, since there exists a singularity in (5.27) for $y_{\bar{x}} = x_1 = 0$ we cannot construct a slave system for η_3 in the proposed form, to overcome this problem, we propose the following slave system:

$$\hat{\eta}_3^{(\alpha)} = \bar{x}_1 \hat{\eta}_2 - d\hat{\eta}_3 \tag{5.29}$$

In order to estimate x_2, we use (5.5), then we have

$$\hat{\eta}_2^{(\alpha)} = K_{2,0}(\eta_2 - \hat{\eta}_2) + K_{2,1} D^{(-\alpha)}(\eta_2 - \hat{\eta}_2) + K_{2,2} D^{(-2\alpha)}(\eta_2 - \hat{\eta}_2) \tag{5.30}$$

$K_{2,j}, j = 0, 1, 2$ are selected such that $P(s^\alpha) = s^{3\alpha} + K_{2,0} s^{2\alpha} + K_{2,1} s^\alpha + K_{2,2}$ be stable.

Looking to avoid time derivatives of the output consider $\gamma_2 = \hat{\eta}_2 - \dfrac{K_{2,0}}{a} y_1$ as an auxiliary variable, now we take (5.28) and after some algebraic manipulations we have

$$\gamma_2^{(\alpha)} = K_{2,0}(y_1 - (\gamma_2 + \frac{K_{2,0}}{a} y_1)) + \frac{K_{2,1}}{a} y_1 + K_{2,1} D^{(-\alpha)}(y_1 - (\gamma_2 + \frac{K_{2,0}}{a} y_1) + K_{2,0} y_1)$$
$$+ \frac{K_{2,2}}{a} D^{(-\alpha)} y_1 + K_{2,2} D^{(-2\alpha)}(y_1 - (\gamma_2 + \frac{K_{2,0}}{a} y_1)) \tag{5.31}$$

We consider equations (5.31) and (5.29), then it is possible to obtain the estimates η_2 and η_3 from the following relations:

$$\hat{\eta}_2 = \frac{K_{2,0}}{a} y_1 + \gamma_2$$

$$\hat{\eta}_3^{(\alpha)} = y_1 \hat{\eta}_2 - d\hat{\eta}_3 \tag{5.32}$$

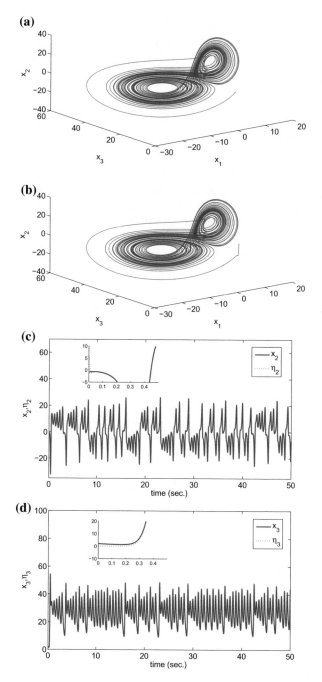

Fig. 5.2 Synchronization of the fractional-order Lorenz system, **a** $x_1 - x_3 - x_2$ space **b** $x_1 - \hat{\eta}_3 - \hat{\eta}_2$ space **c** and **d** shows the convergence of the estimates $\hat{\eta}_2$ and $\hat{\eta}_3$ to the states x_2 and x_3

Numerical simulations were performed with the following parameters: $a = 10$, $b = 28$, $c = -2$, $d = \frac{8}{3}$, $\alpha = 0.96$ with the initial conditions $x_0 = [0.5, -0.9, 2]^T$ and the initial conditions for the slave system are given by $\hat{\eta}_{20} = 10$, $\hat{\eta}_{30} = 20$. The convergence of the estimates to the true signals is shown in Fig. 5.2.

5.4   Conclusion

In this chapter, it was introduced, a new fractional model-free $PI^{r\alpha}$-reduced-order observer inspired in the new concept of fractional Algebraic Observability (FAO), we applied the results to a Rössler hyperchaotic fractional-order system and Lorenz fractional-order system, however, this technique can be applied to another class of systems, which satisfy the properties of Theorem 5.1. Some numerical simulations have illustrated the effectiveness of the suggested approach.

References

1. A. A. Kilbas, H. M. Srivastava and J. J. Trujillo, *Theory and applications of fractional differential equations*, Elsevier B.V., 2006.
2. R. Caponetto, G. Dongola, L. Fortuna and I. Petrás, *Fractional order systems modeling and control applications*, World Scientific, 2010.
3. A. Fradkov, *Cybernetical physics: from control of chaos to Quantum control*, Berlin: Springer, 2007.
4. C. Li and G. Chen, *Chaos and hyperchaos in fractional-order Rössler equations*, Physica A, vol. 341, pp. 55–61, 2004.

Chapter 6
Estimators for a Class of Commensurate Fractional-Order Systems with Caputo Derivative

6.1 Introduction

Using some concepts of observability, based on algebraic properties of fractional-order systems, we can synthesize observers for nonlinear fractional-order systems. The technique used in this chapter is based on Fractional Algebraic Observability property [1, 2]. The former verifies whether a given state of a system can be estimated from a function that depends on the output, input and their finite number of fractional-order derivatives, i.e., if state is reconstructible from output and input measurements. The methodology proposed consist in finding a canonical form for the original system, this is obtained through a mapping given by the output of the system and its successive fractional order derivatives. Then, we can design an observer to estimate the state of the transformed system, so-called Fractional Generalized Observability Canonical Form [1]. Finally, from an inverse mapping, obtain the estimates of the original state.

In general, the structure of the dynamics of the observer is composed of a copy of the system and a correction term. In this chapter, we propose two fractional-order nonlinear observers for a class of commensurate fractional-order systems in an algebraic setting: a reduced-order observer and a Luenberger like observer. The first one is given when only part of the state is necessary to estimate, this is a more natural way to design an observer, where the main feature is that we do not need to know beforehand the system itself, thus the reconstruction of the state should be given only using the system's output. On the other hand, Luenberger observer needs a full copy of the system and requires more fractional-order integrators at implementation level. It should be noted that in case of the reduced order observer is necessary to construct an observer for each unknown variable. Finally, a comparison between two observers is performed with two different numerical examples: a linear mechanical oscillator with an integer and a fractional-order damping [3], and a nonlinear fractional order Duffing System. The main contribution in this chapter is to show a flexible methodology for the observation problem from very simple algebraic techniques.

© Springer Nature Switzerland AG 2018
R. Martínez-Guerra and C. A. Pérez-Pinacho, *Advances in Synchronization of Coupled Fractional Order Systems*, Understanding Complex Systems, https://doi.org/10.1007/978-3-319-93946-9_6

The rest of this chapter is organized as follows: the problem formulation is given in Sect. 6.2, the design and convergence of the fractional-reduced-order observer is detailed in Sect. 6.3, in Sect. 6.4 we show some numerical simulations and finally some concluding remarks are stated in Sect. 6.5.

6.2 Problem Formulation

We take the initial condition problem for an autonomous fractional-order nonlinear system:

$$x^{(\alpha)} = f(x, u), \quad x(0) = x_0,$$
$$\bar{y} = h(\bar{x}),$$
$$(6.1)$$

where $x \in \Omega \subset \mathbb{R}^n$, $f : \Omega \to \mathbb{R}^n$ is a Lipschitz continuous function,[1] with $x_0 \in \Omega \subset \mathbb{R}^n$, $u \in \mathbb{R}^m$ denotes the input, in this case, y denotes the output of the system (the physical measurement that we can obtain), $\bar{x} \in \mathbb{R}^p$ represents the states that we can observe (known states), $h : \mathbb{R}^p \to \mathbb{R}^q$ is a continuous function and $1 \le p < n$. Quite often, due to costs and type of variables, we only have few available state variables to be physically measured and the other state variables should be estimated. From now on, consider the case, where $q = m = 1$, which naturally appears in practical situations. Let $\alpha = [\alpha_1, \alpha_2, \ldots, \alpha_n]^T$, the system (6.1) is called commensurate order system if $\alpha = \alpha_1 = \cdots = \alpha_n$, otherwise it is an incommensurate order system. We consider a commensurate order system with $0 < \alpha < 1$.

Here, observability is regarded as a property to infer the system states x_i, $1 \le i \le n$ from the knowledge of its output $\bar{y}$, input u and a finite number of it fractional-order derivatives. This property is similar to the one used in nonlinear integer-order systems (see [1, 2]). Thus, system (6.1) is called observable if it satisfies FAO (see Definition 2.4).

Inspired by the Theorem of differential primitive element [2], we propose the concept of fractional Picard–Vessiot (PV) (see Definition 2.5), and we can get the FGOCF by the Lemma 2.1.

Remark 6.1 We can take the nonlinear system (6.1) to the FGOCF by means of

$$\eta = (\eta_1 \ldots \eta_n)^T = \Phi(x), \qquad (6.2)$$

where $\Phi : \mathbb{R}^n \to \mathbb{R}^n$ is a nonlinear mapping obtained from (2.22). Thus, from the fractional differential primitive element (the available output of system (6.1)), we can obtain its FGOCF.

At this point, the observation problem setting is clear. If we obtain an estimation of the unknown fractional-order derivatives of the output (2.22) and due to original system (6.1) is observable in the sense of FAO condition, the observation problem of

---

[1]This assures the unique solution, see [4].

system (6.1) is solved if we design an observer for this transformed system (2.23). Roughly speaking, this observation problem becomes an estimation problem of the unknown fractional-order derivatives of the output.

6.3 Main Result

6.3.1 Fractional-Reduced-Order Observer (FROO)

Consider the system given by (2.23), without loss of generality assume $u = 0$, we will separate in two dynamical systems with states $\eta_1 \in \mathbb{R}$ and $\hat{\eta}^T = (\hat{\eta}_2, \ldots, \hat{\eta}_n) \in \mathbb{R}^{n-1}$ respectively, the whole state is grouped as $\eta^T = (\eta_1, \hat{\eta}^T)$, the first system describe the known state and the second represents unknown states to be estimated, then the system (2.23) can be rewritten as

$$
\begin{aligned}
\bar{\eta}_1^{(\alpha)} &= \hat{\eta}_2, \\
\hat{\eta}^{(\alpha)} &= \Delta(\eta_1, \hat{\eta}), \\
\bar{y} &= \eta_1,
\end{aligned}
\tag{6.3}
$$

where $f_\eta^T(\eta_1, \hat{\eta}) = (\eta_1, \Delta^T(\bar{\eta}_1, \hat{\eta}))$ and $\Delta : \mathbb{R} \times \mathbb{R}^{n-1} \to \mathbb{R}^{n-1}$. If we assume that the components of unknown state vector $\hat{\eta}$ are FAO (in fact it is, see (2.22)), then we can describe our problem in the following way. Let us consider the reduced-order system

$$
\begin{aligned}
\hat{\eta}_i^{(\alpha)} &= \Delta_i(\eta_1, \hat{\eta}), \\
\hat{\eta}_i &= \phi_i\left(\bar{y}, \bar{y}^{(\alpha)}, \mathscr{D}^{2\alpha}\bar{y}, \ldots, \mathscr{D}^{r\alpha}\bar{y}\right),
\end{aligned}
\tag{6.4}
$$

for $2 \le i \le n$, where $\hat{\eta}_i$ is the ith component of the state vector $\hat{\eta}$ that satisfy FAO property and consider the unknown dynamics

$$
\hat{\eta}_i^{(\alpha)} = \bar{\Delta}_i(\bar{\eta}_1, \hat{\eta}),
\tag{6.5}
$$

where $\bar{\Delta}_i(\eta_1, \hat{\eta})$ is an unknown dynamics related to system (6.4). In the next Lemma, we propose a fractional-order dynamical system to estimate these unknown dynamics.

Lemma 6.1 *Let system* (6.1) *be transformable into a FGOCF, assume output $\bar{y}$ and its fractional derivatives are bounded, i.e., $\exists M_i \in \mathbb{R}^+$ such that $\|\eta_i^{(\alpha)}\| \le M_i$. And consider the reduced-order observer for the fractional derivatives of output:*

$$
\hat{\eta}_i^{(\alpha)} = k_{\hat{\eta}_i}(\eta_i - \hat{\eta}_i),
\tag{6.6}
$$

with a constant gain $k_{\hat{\eta}_i} > 0$, *for* $2 \le i \le n$. *Then, system*

$$
\begin{aligned}
\gamma_{\hat{\eta}_i}^{(\alpha)} &= -k_{\hat{\eta}_i} \left(\gamma_{\hat{\eta}_i} + k_{\hat{\eta}_i} \bar{\eta}_{i-1} \right), \quad \gamma_{\hat{\eta}_i}(0) = \gamma_{\hat{\eta}_i 0} \in \mathbb{R}, \\
\hat{\eta}_i &= \gamma_{\hat{\eta}_i} + k_{\hat{\eta}_i} \bar{\eta}_{i-1},
\end{aligned}
\tag{6.7}
$$

is a reduced-order observer for FGOCF. Moreover, estimation error $\varepsilon_i = \eta_i - \hat{\eta}_i$ *asymptotically converges to* $\bar{B}_{\rho_i}(0) = \{\varepsilon_i \in \mathbb{R} \mid |\varepsilon_i| \le \rho_i\}$, *with* $\rho_i := N_i / k_{\hat{\eta}_i}$.

Proof Consider the fractional reduced-order observer for unknown states (derivatives of output) in the FGOCF:

$$
\hat{\eta}_i^{(\alpha)} = k_{\hat{\eta}_i} \eta_i - k_{\hat{\eta}_i} \hat{\eta}_i,
\tag{6.8}
$$

for $2 \le i \le n$, due to η_i is algebraically observable, i.e.,

$$
\eta_i = \bar{\eta}_{i-1}^{(\alpha)},
\tag{6.9}
$$

then, system (6.8) is equivalent to

$$
\hat{\eta}_i^{(\alpha)} = k_{\hat{\eta}_i} \bar{\eta}_{i-1}^{(\alpha)} - k_{\hat{\eta}_i} \hat{\eta}_i.
\tag{6.10}
$$

Define an auxiliary variable $\gamma_{\hat{\eta}_i}$ as follows

$$
\gamma_{\hat{\eta}_i} = -k_{\hat{\eta}_i} \bar{\eta}_{i-1} + \hat{\eta}_i,
$$

thus,

$$
\hat{\eta}_i = \gamma_{\hat{\eta}_1} + k_{\hat{\eta}_i} \bar{\eta}_{i-1},
\tag{6.11}
$$

obtaining its fractional derivative of order α, we have

$$
\hat{\eta}_i^{(\alpha)} = \gamma_{\hat{\eta}_i}^{(\alpha)} + k_{\hat{\eta}_i} \bar{\eta}_{i-1}^{(\alpha)}.
\tag{6.12}
$$

Considering (6.11) and (6.12) in (6.10), it is obtained

$$
\begin{aligned}
\gamma_{\hat{\eta}_i}^{(\alpha)} &= -k_{\hat{\eta}_i} (\gamma_{\hat{\eta}_i} + k_{\hat{\eta}_i} \bar{\eta}_{i-1}), \\
\hat{\eta}_i &= \gamma_{\hat{\eta}_i} + k_{\hat{\eta}_i} \bar{\eta}_{i-1}.
\end{aligned}
\tag{6.13}
$$

On the other hand, define the observation error for the ith observer as follows:

$$
\varepsilon_i = \eta_i - \hat{\eta}_i = \eta_i - \gamma_{\hat{\eta}_i} - k_{\hat{\eta}_i} \bar{\eta}_{i-1},
\tag{6.14}
$$

obtaining the fractional derivative of (6.14) gives

$$
\varepsilon_i^{(\alpha)} = \eta_i^{(\alpha)} - \gamma_{\hat{\eta}_i}^{(\alpha)} - k_{\hat{\eta}_i} \bar{\eta}_{i-1}^{(\alpha)},
\tag{6.15}
$$

from the algebraic observability property and (6.13) we obtain

$$\varepsilon_i^{(\alpha)} = \eta_i^{(\alpha)} + k_{\hat{\eta}_i}\gamma_{\hat{\eta}_i} + k_{\hat{\eta}_i}^2 \bar{\eta}_{i-1} - k_{\hat{\eta}_i}\eta_i, \tag{6.16}$$

by adding the null term $k_{\hat{\eta}_i}\hat{\eta}_i - k_{\hat{\eta}_i}\hat{\eta}_i$ yields to the dynamics of the estimation error given by

$$\varepsilon_i^{(\alpha)} + k_{\hat{\eta}_i}\varepsilon_i = \eta_i^{(\alpha)} = \Delta_i(t). \tag{6.17}$$

There exists a unique solution for the system (6.17), due to $\Delta_i(t) - k_{\hat{\eta}_i}\varepsilon_i(t)$ is a Lipschitz continuous function on ε_i and uniformly in $\Delta_i(t)$.[2] The solution of (6.17) is taken from [4, 5], and is given by

$$\begin{aligned}\varepsilon_i(t) &= \varepsilon_{i0}E_{\alpha,1}(-k_{\hat{\eta}_i}t^\alpha) \\ &+ \int_0^t (t-\tau)^{\alpha-1}E_{\alpha,\alpha}(k_{\hat{\eta}_i}(t-\tau)^\alpha)\Delta_i(\tau)d\tau,\end{aligned} \tag{6.18}$$

where $\varepsilon_i(0) = \varepsilon_{i0}$. Using Triangle and Cauchy–Schwarz inequalities and the fact that $\|\Delta_i(t)\| \le M_i$

$$\begin{aligned}|\varepsilon_i(t)| &\le |\varepsilon_{i0}E_{\alpha,1}(-k_{\hat{\eta}_i}t^\alpha)| \\ &+ M_i \int_0^t |(t-\tau)^{\alpha-1}E_{\alpha,\alpha}(-k_{\hat{\eta}_i}(t-\tau)^\alpha)|d\tau.\end{aligned}$$

The functions $(t-\tau)^{\alpha-1}E_{\alpha,\alpha}(-k_{\hat{\eta}_i}(t-\tau)^\alpha)$ and $E_\alpha(-k_{\hat{\eta}_i}t^\alpha)$ are not negative due to Property 2 of Mittag–Leffler function and $k_{\hat{\eta}_i} > 0$

$$\begin{aligned}|\varepsilon_i(t)| &\le |\varepsilon_{i0}|E_{\alpha,1}(-k_{\hat{\eta}_i}t^\alpha) \\ &+ M_i \int_0^t (t-\tau)^{\alpha-1}E_{\alpha,\alpha}(-k_{\hat{\eta}_i}(t-\tau)^\alpha)d\tau.\end{aligned}$$

Using Property 1 of Mittag–Leffler function

$$|\varepsilon_i(t)| \le |\varepsilon_{i0}|E_{\alpha,1}(-k_{\hat{\eta}_i}t^\alpha) + M_i t^\alpha E_{\alpha,\alpha+1}(-k_{\hat{\eta}_i}t^\alpha).$$

If $t \to \infty$, by using (2.5) from Theorem 2.1 with $\mu = 3\pi\alpha/4$ and due to $k_{\hat{\eta}_i} > 0$, then

$$\begin{aligned}\lim_{t\to\infty}|\varepsilon_i(t)| &\le |\varepsilon_{i0}| \lim_{t\to\infty} E_{\alpha,1}(-k_{\hat{\eta}_i}t^\alpha) \\ &+ M_i \lim_{t\to\infty} t^\alpha E_{\alpha,\alpha+1}(-k_{\hat{\eta}_i}t^\alpha) = \frac{M_i}{k_{\hat{\eta}_i}}.\end{aligned}$$

Remark 6.2 The use of the reduced-order observer in (6.7) should be done in an iterative way, that is, first obtain an estimation for the first unknown variable then for the second and so on. This procedure gives a bank of reduced-order observers and

---

[2]Equation (6.17) is nonautonomous, but the Lipschitz condition assures a unique solution [4, 5].

Fig. 6.1 Observation scheme for the fractional-order reduced-order observer

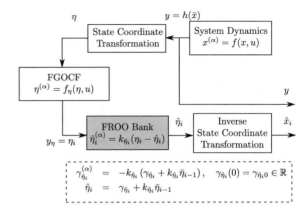

reduce the use of fractional-order integrators. Notice that this type of observer does not need a copy of system (2.23). The former becomes an important issue compare to estimation with a classical Luenberger observer.

Remark 6.3 A reduced-order observer of the form (6.6) can be designed directly for system (6.1) (see [2]). The drawback of that approach is that state variables are expressed in terms of the unknown fractional derivatives of η_i, thus it is necessary to introduce an artificial variable which is more complicated variable than $\gamma_{\hat{\eta}_i}$ in order to avoid the use of these unknown derivatives. As a result a complicated reduced-order observer, for each state of the original system, is obtained.

Corollary 6.1 *Let a nonlinear fractional-order system* (6.1) *to be transformable into a FGOCF, assume inverse transformation* $\Phi^{-1}(\cdot)$ *in* (6.2) *exist. Then, the dynamical system* (6.7) *along with:*

$$x = \Phi^{-1}(\eta),$$

constitute an stable observer for the nonlinear fractional-order system (6.1).

Finally, Corollary 6.1 is depicted in Fig. 6.1 where full observation setting is shown.

6.3.2 Fractional-Order Luenberger Observer (FLO)

Consider system (6.1) in its FGOCF (2.23), we can rewrite this system in the Brunovsky's canonical form:

$$
\begin{aligned}
\eta^{(\alpha)} &= A\eta - B\mathcal{T}(\eta, \bar{u}), \\
\bar{y} &= C\eta,
\end{aligned}
\tag{6.19}
$$

where matrices $A_{n\times n}$, $B_{n\times 1}$ and $C_{1\times n}$ are given by

$$A = \begin{pmatrix} 0 & 1 & \dots & 0 \\ \vdots & \vdots & \ddots & 1 \\ 0 & 0 & \dots & 0 \end{pmatrix}, \quad B = \begin{pmatrix} 0 \\ \vdots \\ 0 \\ 1 \end{pmatrix}, \quad C = \begin{pmatrix} 1 & 0 & \dots & 0 \end{pmatrix}.$$

On the other hand, let us consider a copy of system (6.19) with a weighted correction term with gain $L = (l_1 \ l_2 \ \dots \ l_n)^T \in \mathbb{R}^n$ as the nonlinear fractional-order Luenberger observer:

$$\hat{\eta}^{(\alpha)} = A\hat{\eta} - B\mathcal{T}(\hat{\eta}, \bar{u}) + L(\bar{y} - \hat{y}). \tag{6.20}$$
$$\hat{y} = C\hat{\eta}.$$

Now, we are in position to establish the next convergence analysis for estimation error, $\varepsilon := \eta - \hat{\eta}$, $\varepsilon = (\varepsilon_1 \cdots \varepsilon_n)^T \in \mathbb{R}^n$, resumed in the following proposition.

Proposition 6.1 *Let system* (6.1) *be transformable into a FGOCF and consider the full state observer*

$$\hat{\eta}^{(\alpha)} = A\hat{\eta} - B\bar{\mathcal{T}}(\hat{\eta}, \bar{u}) + L(\bar{y} - \hat{y}),$$
$$\hat{y} = C\hat{\eta},$$

with a constant gain vector L. Assume

- $\bar{A} = (A - LC)$ *is a Hurwitz matrix with* $\lambda_i(\bar{A}) \neq \lambda_j(\bar{A})$, $1 \leq i, j \leq n$, $i \neq j$,
- $|\mathcal{T}(\eta, \bar{u}) - \mathcal{T}(\hat{\eta}, \bar{u})| \leq N \in \mathbb{R}^+$,

- *Permutation matrix* $P = \begin{pmatrix} 0 & & 1 \\ & \cdot^{\cdot^{\cdot}} & \\ 1 & & 0 \end{pmatrix}$.

Then, estimation error $\varepsilon = \eta - \hat{\eta}$ *is asymptotically bounded, i.e.,* $|\varepsilon| \leq N\|PV^{-1}\|_\infty$ $\max\{|1/\lambda_1|, \dots, |1/\lambda_n|\}$ *as* $t \to \infty$, *where* V^{-1} *is the inverse Vandermonde matrix of* $\bar{A}$.

Proof Take the fractional derivative of estimation error $\varepsilon^{(\alpha)} = \eta^{(\alpha)} - \hat{\eta}^{(\alpha)}$. Using (6.19) and (6.20), we have

$$\varepsilon^{(\alpha)} = (A - LC)\varepsilon - B \underbrace{(\mathcal{T}(\eta, \bar{u}) - \mathcal{T}(\hat{\eta}, \bar{u}))}_{\bar{\mathcal{T}}(\hat{\eta}, \eta, \bar{u})}. \tag{6.21}$$

Choosing gain matrix L such that $\bar{A} = (A - LC)$ is a Hurwitz matrix with distinct eigenvalues $\lambda_i(\bar{A}) \neq \lambda_j(\bar{A})$, $1 \leq i, j \leq n$, $i \neq j$. We can obtain a modal decomposition of system (6.21) using a similarity transformation involving Vandermonde matrix V (see Property 3, assuming its inverse V^{-1} exists) and matrix P

$$\dot{\varepsilon}_V = PV\varepsilon, \tag{6.22}$$

and the inverse relation

$$\varepsilon = PV^{-1}\varepsilon_V. \tag{6.23}$$

From (6.21), (6.22), and (6.23) we obtain the next decomposition

$$\varepsilon_V^{(\alpha)} = VP\varepsilon^{(\alpha)} = \underbrace{VP\bar{A}PV^{-1}}_{D}\varepsilon_V - VPB\bar{\mathscr{T}}(\eta, \hat{\eta}, \bar{u}),$$

$$= D\varepsilon_V - \bar{B}\bar{\mathscr{T}}(\eta, \hat{\eta}, \bar{u}),$$

where $\bar{B} = (1 \cdots 1)^T \in \mathbb{R}^n$. Hence,

$$\varepsilon_{V_i}^{(\alpha)} = \lambda_i \varepsilon_{V_i} - \bar{\mathscr{T}}(\eta, \hat{\eta}, \bar{u}), \tag{6.24}$$

the solution for the Eq. (6.24) is the following

$$\varepsilon_{V_i} = \varepsilon_{V_i 0} E_{\alpha, i}(\lambda_i t^{(\alpha)}) +$$

$$- \int_0^t (t - \tau)^{\alpha - 1} E_{\alpha, \alpha}(\lambda_i (t - \tau)^\alpha) \bar{\mathscr{T}}(\eta(\tau), \hat{\eta}(\tau), \bar{u}(\tau)) d\tau.$$

Using the same arguments as in proof of Lemma 6.1 and assuming $\bar{\mathscr{T}}(\cdot)$ is a bounded function $|\bar{\mathscr{T}}(\cdot)| \leq N$, the error $\varepsilon_{V_i 0}$ asymptotically converges to

$$\lim_{t \to \infty} |\varepsilon_{V_i}(t)| \leq |\varepsilon_{V_i 0}| \lim_{t \to \infty} E_{\alpha, 1}(\lambda_i t^\alpha)$$

$$+ N \lim_{t \to \infty} t^\alpha E_{\alpha, \alpha + 1}(\lambda_i t^\alpha) = -\frac{N}{\lambda_i}.$$

Finally, the next estimate fulfills[3]:

$$\lim_{t \to \infty} \| \varepsilon_V(t) \|_\infty \leq N \max \left\{ \left| \frac{1}{\lambda_1} \right|, \ldots, \left| \frac{1}{\lambda_n} \right| \right\}.$$

And consequently from (6.23),

$$\lim_{t \to \infty} \|\varepsilon\|_\infty \leq N \| PV^{-1} \|_\infty \max \left\{ \left| \frac{1}{\lambda_1} \right|, \ldots, \left| \frac{1}{\lambda_n} \right| \right\}.$$

---

[3]For $z = (z_1, \ldots, z_n)^T \in \mathbb{R}^n$, $\| x \|_\infty := \max \{|x_1|, \ldots, |x_n|\}$.

6.4 Numerical Examples

6.4.1 Fractional-Order Linear Mechanical Oscillator

Consider the following forced fractional-order linear mechanical system composed of an oscillator with classical fluid damping and frictious fractional damping [3]:

$$\ddot{x} + a\dot{x} + bd^{1/2}x + cx = u. \tag{6.25}$$

The measured output y is position x, which can be physically measured, contrarily to $d^{1/2}x$, $\dot{x}$ and $d^{1/2}\dot{x}$. This system has a commensurate fractional order of $\alpha = 1/2$, such that, in state space form we can write system (6.25) as follows:

$$\begin{aligned} x_j^{(\alpha)} &= x_{j+1}, \quad 1 \le j \le 3, \\ x_4^{(\alpha)} &= -ax_3 - bx_2 - cx_1 + u, \end{aligned} \tag{6.26}$$

with $x(0) = (-2 \ -1 \ 2 \ 0.5)^T$. Take $y = x_1$ as the fractional differential primitive element, thus FAO condition is fulfilled

$$y^{([\ell-1]\alpha)} - x_\ell = 0, \quad 1 \le \ell \le 4,$$

Hence, system (6.26) can be transformed into

$$\begin{aligned} \eta_j^{(\alpha)} &= \eta_{j+1} \quad 1 \le j \le 3, \\ \eta_4^{(\alpha)} &= -a\eta_3 - b\eta_4 - c\eta_1 + u. \end{aligned}$$

Now we want to estimate η_2, η_3, η_4 from measurements of η_1. From Lemma 6.1, it is not hard to obtain the following state observer for the FGOCF, the bank of reduced-order observers is given by

$$\begin{aligned} \gamma_{\hat{\eta}_2}^{(\alpha)} &= -K_{\hat{\eta}_2}\gamma_{\hat{\eta}_2} - K_{\hat{\eta}_2}^2 y_\eta \\ \gamma_{\hat{\eta}_i}^{(\alpha)} &= -K_{\hat{\eta}_i}\gamma_{\hat{\eta}_i} - K_{\hat{\eta}_i}^2 \hat{\eta}_{i-1} \quad i = 3, 4 \end{aligned} \tag{6.27}$$

with

$$\begin{aligned} \hat{\eta}_2 &= \gamma_{\hat{\eta}_2} + K_{\hat{\eta}_2} y_\eta \\ \hat{\eta}_i &= \gamma_{\hat{\eta}_i} + K_{\hat{\eta}_i}\hat{\eta}_{i-1} \quad i = 3, 4. \end{aligned}$$

Numerical simulations corresponding to Luenberger observer for the system (6.26), and the bank of reduced-order observers (6.27) are shown in Figs. 6.2 and 6.3. Clearly, both observers estimate the state of the original system. However, to com-

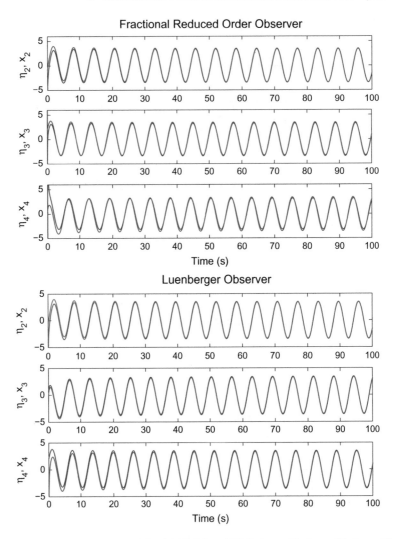

Fig. 6.2 Estimation of system (6.26) with FROO and FLO, $k_{\hat{\eta}_2} = 10$, $k_{\hat{\eta}_2} = 15$, $k_{\hat{\eta}_4} = 20$ and $L = (10\ 15\ 20\ 1)^T$ (top to bottom, respectively)

pare the performance of both observers, consider the Integral Squared Error (ISE) index[4] with $\kappa = 100$, $T_2 - T_1 = 0.5$ s.

---

[4]Due to lack of space we only compare the performance of both observers based on the same observed trajectories with:

$$ISE = \int_{T_1}^{T_2} (\kappa \varepsilon_i(\tau))^2 \, d\tau.$$

This index tolerates small errors for long periods and penalize long errors.

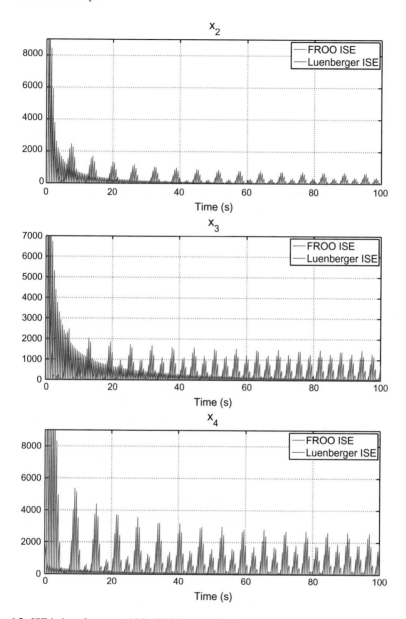

Fig. 6.3 ISE index of system (6.26): FROO versus FLO

6.4.2 Duffing System

Consider the following fractional nonlinear Duffing Systems [1]:

$$x_1^{(\alpha)} = x_2 \tag{6.28}$$
$$x_2^{(\alpha)} = x_1 - x_1^3 - \beta x_2 + \delta cos(\omega t),$$

where $\beta = 0.15$ is the damping coefficient, $\delta = 1.3$ is the amplitude of the forcing function and $\omega = 1$ is the forcing frequency and the initial conditions $x(0) = (1 \ 1)^T$, and $y = x_1$ are the measures the horizontal displacement. From Lemma 6.1, the reduced order observer for Duffing system is given by

$$\gamma_{\hat{\eta}_2}^{(\alpha)} = -K_{\hat{\eta}_2}\gamma_{\hat{\eta}_2} - K_{\hat{\eta}_2}^2 y$$
$$\hat{\eta}_2 = \gamma_{\hat{\eta}_2} + K_{\hat{\eta}_2} y$$

In the Figs. 6.4 and 6.5, we show the performance of estimating state x_2 of duffing system.

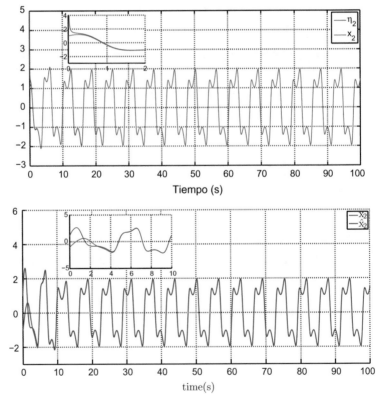

Fig. 6.4 Estimation of system (6.28) with FROO and FLO, $k_{\hat{\eta}_2} = 200$ and $L = (11 \ 12)^T$ (top to bottom respectively)

Fig. 6.5 ISE index of system (6.28): FROO versus FLO

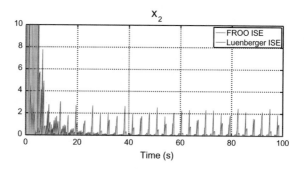

6.5 Conclusions

In this chapter, we have proposed two type of observers (e.g., FROO and FLO) for commensurate fractional-order nonlinear systems, where FAO condition is fundamental to determine whether the system is observable from the knowledge of the input and output and moreover. The FROO posses an acceptable performance without the knowledge of the system dynamics.

References

1. Rafael Martínez-Guerra, Claudia A. Pérez-Pinacho & Gian Carlo Gómez-Cortés. Synchronization of Integral and Fractional Order Chaotic Systems: A Differential Algebraic and Differential Geometric Approach With Selected Applications in Real-Time, Springer 2015.
2. Rafael Martínez-Guerra & Christopher D. Cruz-Ancona. Algorithms of Estimation for Nonlinear Systems: A Differential and Algebraic Viewpoint, Springer 2017.
3. D. Matignon & B. d'Andrea-Novel. Observer-based controllers for fractional differential systems, Proc. of the 36th CDC, San Diego, Ca. USA, 5: 4967–4972, (December) 1997.
4. Kilbas, A., Srivastava, H., and Trujillo, J., Theory and Applications of Fractional Differential Equations., Elsevier B. V., (2006).
5. Kai Diethelm. The Analysis of Fractional Differential Equations: An Application-Oriented Exposition Using Differential Operators of Caputo Type, Lecture Notes in Mathematics, Springer 2004.

Chapter 7
Generalized Multi-synchronization of Fractional Order Liouvillian Chaotic Systems Using Fractional Dynamical Controller

7.1 Introduction

The attempt to understand the synchronize of a pair of systems have been extended to the study of a more complex problem involving multiple systems, of course, motivated by problems in the integer order case where synchronization is observed such as rendezvous, formation control, flocking and schooling, attitude alignment, sensor networks, distributed computing, consensus, and complex networks in general [1–8]. In the fractional order case, we can mention as related work from a control theory perspective: A multi-synchronization scheme for identical systems in a ring connection with unidirectional and bidirectional coupling [9]. In [10], a pinning synchronization problem is presented for a network of systems with Lipschitz-type nonlinearities and unidirectional configuration. In terms of generalized synchronization, a modification of active control is given in [11] where it is considered the case of interaction between multiple slave systems. As in general and in all aforementioned contributions, proving stability of the origin of synchronization error, or synchronization manifold if possible, is the main task in synchronization problems. It is well known that in the case of interacting systems with identical dynamics, there exists a trivial synchronization algebraic manifold (e.g., CS problem). However, for systems with strictly different dynamics, the synchronization manifold is not trivial or it does not necessarily exist. Therefore, it is unclear whether this type of interacting systems can synchronize [4]. Thus, differences in the dynamical structure of the systems play an interesting role, the study of the synchronization error (synchronization manifold) is more challenging than the case of identical systems. Besides, the individual dynamical structure of groups of interacting systems are different in most cases [1, 6, 8]. For synchronization of strictly different fractional order systems, we can mention the work in [11–15].

© Springer Nature Switzerland AG 2018
R. Martínez-Guerra and C. A. Pérez-Pinacho, *Advances in Synchronization of Coupled Fractional Order Systems*, Understanding Complex Systems,
https://doi.org/10.1007/978-3-319-93946-9_7

In recent years, active control has been a popular technique used in chaos synchronization. These controllers use a statical control signal and for each first-order fractional differential equation [16], this suggests the search of appropriate signals to obtain a stable error dynamics. This motivates us to propose, as part of our work, a reduced number of fractional dynamical control laws is able to stabilize the origin of error dynamics with less effort in searching the control signals.

In this chapter, we propose a methodology to synchronize a class of multiple chaotic decoupled nonlinear fractional order systems, where systems are not necessarily identical and is sufficient to construct a fractional differential primitive element based on output of each system that generates a family of transformations. Thus, systems are carried out to a Multi-Output Fractional Generalized Observability Canonical Form (MFGOCF). It is worth mentioning that the family of fractional differential primitive elements is given in a natural manner as a linear combination of the known state and inputs of the systems and their finite number of fractional derivatives. Finally, from this coordinate transformation, we can synchronize multiple decoupled families of chaotic systems in a master–slave configuration. This type of synchronization is introduced as Fractional Generalized Multi-Synchronization (FGMS). Moreover, it is also considered the case of complex interaction between slave systems as a natural extension of the former result, as it will be shown that any type of interplay between slave systems can be considered and still needs to obtain synchronization error convergence to the origin. The main key ingredient is to find canonical forms for the original systems that are synchronized by means of a family of fractional dynamical control signals through a chain of fractional integrators.

There are several contributions in this chapter: First, a methodology to synchronize multiple decoupled families of commensurate fractional order Liouvillian systems where systems need not to be identical. This is given for a master–slave configuration considering also the case of complex interaction between slave systems, both with full access information of the master systems' dynamics.

Second, an explicit construction of the mapping that relates the trajectories of the master and slaves systems by means of the fractional differential primitive element can be obtained. We use the special characteristic of a class of systems so-called fractional Liouvillian systems. It should be noted that first component of these transformations is a fractional order integral of the fractional differential primitive element.

Third, a fractional dynamical control law able to achieve FGMS for all slave systems is given. A natural extension is given for the case of complex interaction between slave systems, this consist in adding diffusive coupling terms (weighted difference between the states of a slave system and the states of the neighbors slave systems) in the dynamical control law. It is worth mentioning that FGMS is given regardless the type interplay between slave systems.

Finally, we introduced definitions related to the concept of fractional Liouvillian algebraic observability, Picard–Vessiot systems, and the concept of Fractional Generalized Multi-Synchronization for fractional order systems.

The reminder of this chapter is organized as follows: The formulation of the problem and main result regarding to generalized synchronization of multiple decoupled fractional order systems is stated in Sect. 7.2. In Sect. 7.3, the extension for the case of complex interaction between slave systems is given. Some examples are provided in Sect. 7.4 with two different fractional order Liouvillian chaotic systems. Finally, in Sect. 7.5, some concluding remarks are given.

7.2 Problem Formulation

In this section, we will see that the synchronization of a class of multiple decoupled nonlinear fractional order systems is reduced to a problem of multiple fractional generalized synchronization where it is sufficient to know the output of each system to generate a family of transformations which gives us the possibility to synchronize multiple chaotic systems, these transformations are obtained from a family of outputs given by $\bar{y}_j = I^\alpha y_j$ with $1 \leq j \leq p$ (p outputs). Let $n_j \geq 0$ be the minimum integers such that $\mathscr{D}^{n_j\alpha}\bar{y}_j$ are analytically dependent on $(\bar{y}_j, \bar{y}_j^{(\alpha)}, \ldots, \mathscr{D}^{[n_j-1]\alpha}\bar{y}_j)$, where $\bar{y}_j = I^\alpha y_j$

$$\bar{H}_j(\bar{y}_j, \bar{y}_j^{(\alpha)}, \ldots, \mathscr{D}^{[n_j-1]\alpha}\bar{y}_j, \mathscr{D}^{n_j\alpha}\bar{y}_j, u_j, u_j^{(\alpha)}, \ldots, \mathscr{D}^{\gamma_j\alpha}u_j) = 0. \qquad (7.1)$$

The system (7.1) can be solved locally as:

$$\mathscr{D}^{n_j\alpha}\bar{y}_j = -\mathscr{L}_j(\bar{y}_j, \bar{y}_j^{(\alpha)}, \ldots, \mathscr{D}^{[n_j-1]\alpha}\bar{y}_j, u_j, u_j^{(\alpha)}, \ldots, \mathscr{D}^{[\gamma_j-1]\alpha}u_j) + \mathscr{D}^{\gamma_j\alpha}u_j$$

Let $\xi_i^{n_j} = \mathscr{D}^{[i-l]\alpha}\bar{y}_j, l = 1, n_1 + 1, n_1 + n_2 + 1, \ldots, n_1 + n_2 + \cdots + n_{p-1} + 1$; $1 \leq i \leq \sum_{1 \leq j \leq p} n_j = n$ where index j gives the $j - th$ system and the n_j's are the so-called index of algebraic observability where each index coincides with the system's dimension. Then, it is possible to achieve a local representation for a set of p decoupled systems, this representation can be seen as a Multi-output Fractional Generalized Observability Canonical Form (MFGOCF):

$$\mathscr{D}^\alpha \xi_1^{n1} = \xi_2^{n1}$$
$$\mathscr{D}^\alpha \xi_2^{n1} = \xi_3^{n1}$$

$$\vdots$$

$$\mathscr{D}^\alpha \xi_{n_1-1}^{n1} = \xi_{n_1}^{n1}$$
$$\mathscr{D}^\alpha \xi_{n_1}^{n1} = -\mathscr{L}_1(\xi_1^{n1}, \xi_2^{n1}, \dots, \xi_{n_1}^{n1}, u_1, u_1^{(\alpha)}, \dots, \mathscr{D}^{[\gamma_1-1]\alpha} u_1) + \mathscr{D}^{\gamma_1 \alpha} u_1$$
$$\mathscr{D}^\alpha \xi_{n_1+1}^{n2} = \xi_{n_1+2}^{n2}$$
$$\mathscr{D}^\alpha \xi_{n_1+2}^{n2} = \xi_{n_1+3}^{n2}$$

$$\vdots$$

$$\mathscr{D}^\alpha \xi_{n_1+n_2-1}^{n2} = \xi_{n_1+n_2}^{n2}$$
$$\mathscr{D}^\alpha \xi_{n_1+n_2}^{n2} = -\mathscr{L}_2(\xi_{n_1+1}^{n2}, \xi_{n_1+2}^{n2}, \dots, \xi_{n_1+n_2}^{n2}, u_2, u_2^{(\alpha)}, \dots, \mathscr{D}^{[\gamma_2-1]\alpha} u_2) + \mathscr{D}^{\gamma_2 \alpha} u_2 \quad (7.2)$$

$$\vdots$$

$$\mathscr{D}^\alpha \xi_{n_1+n_2+\cdots+n_{p-1}+1}^{np} = \xi_{n_1+n_2+\cdots+n_{p-1}+2}^{np}$$
$$\mathscr{D}^\alpha \xi_{n_1+n_2+\cdots+n_{p-1}+2}^{np} = \xi_{n_1+n_2+\cdots+n_{p-1}+3}^{np}$$

$$\vdots$$

$$\mathscr{D}^\alpha \xi_{n_1+n_2+\cdots+n_{p-1}+n_p-1}^{np} = \xi_{n_1+n_2+\cdots+n_{p-1}+n_p}^{np}$$
$$\mathscr{D}^\alpha \xi_{n_1+n_2+\cdots+n_{p-1}+n_p}^{np} = -\mathscr{L}_p(\xi_{n_1+n_2+\cdots+n_{p-1}+1}^{np}, \xi_{n_1+n_2+\cdots+n_{p-1}+2}^{np}, \dots,$$
$$\xi_{n_1+n_2+\cdots+n_{p-1}+n_p}^{np}, u_p, u_p^{(\alpha)}, \dots, \mathscr{D}^{[\gamma_p-1]\alpha} u_p) + \mathscr{D}^{\gamma_p \alpha} u_p$$
$$y_j = \xi_l^{n_j}$$

In a compact form, the new system (7.2) can be represented as:

$$\mathscr{D}^\alpha \xi = \mathscr{A}\xi - \Phi(\mathscr{L}_1, \dots, \mathscr{L}_p) + \bar{\mathscr{U}}(\mathscr{D}^{\gamma_1 \alpha} u_1, \dots, \mathscr{D}^{\gamma_p \alpha} u_p)$$
$$\mathscr{Y} = \mathscr{C}\xi \tag{7.3}$$

where $\xi, \Phi, \bar{\mathscr{U}} \in \mathbb{R}^n$, $\mathscr{A} \in \mathbb{R}^{n \times n}$, $\mathscr{Y} \in \mathbb{R}^p$ and the matrices of (7.3) are defined as follows:

$$\mathscr{A} = \begin{bmatrix} A_1 & & 0 \\ & \ddots & \\ 0 & & A_p \end{bmatrix};$$

$$A_j = \begin{bmatrix} 0 & 1 & 0 & 0 & \cdots & 0 \\ 0 & 0 & 1 & 0 & \cdots & 0 \\ \vdots & \vdots & \vdots & \ddots & \cdots & 0 \\ 0 & 0 & 0 & 0 & 1 & 0 \\ 0 & 0 & 0 & 0 & 0 & 1 \\ 0 & 0 & 0 & 0 & 0 & 0 \end{bmatrix} \quad 1 \le j \le p;$$

$$\Phi(\mathscr{L}_1,\ldots,\mathscr{L}_p) = \begin{bmatrix} \phi_1(\mathscr{L}_1) \\ \phi_2(\mathscr{L}_2) \\ \vdots \\ \phi_p(\mathscr{L}_p) \end{bmatrix};$$

$$\phi_j(\mathscr{L}_j) = \begin{bmatrix} 0 \\ 0 \\ \vdots \\ 0 \\ -\mathscr{L}_j(\xi^{n_j}_{n_1+n_2+\cdots+n_{j-1}+1}, \ldots, \xi^{n_p}_{n_1+n_2+\cdots+n_j}, u_j, u_j^{(\alpha)}, \ldots, \mathscr{D}^{[\gamma_j-1]\alpha} u_j) \end{bmatrix};$$

$$\bar{\mathscr{U}}(\mathscr{D}^{\gamma_1\alpha} u_1, \ldots, \mathscr{D}^{\gamma_p\alpha} u_p) = \begin{bmatrix} \mathscr{U}_1(\mathscr{D}^{\gamma_1\alpha} u_1) \\ \mathscr{U}_2(\mathscr{D}^{\gamma_2\alpha} u_2) \\ \vdots \\ \mathscr{U}_p(\mathscr{D}^{\gamma_p\alpha} u_p) \end{bmatrix}; \quad \mathscr{U}_j(\mathscr{D}^{\gamma_j\alpha} u_j) = \begin{bmatrix} 0 \\ 0 \\ \vdots \\ 0 \\ \mathscr{D}^{\gamma_j\alpha} u_j \end{bmatrix};$$

$$\mathscr{C} = \begin{bmatrix} C_1 & & 0 \\ & \ddots & \\ 0 & & C_p \end{bmatrix}; \quad C_j = \begin{bmatrix} 1 & 0 & \cdots & 0 \end{bmatrix}.$$

Now, consider the following family of chaotic nonlinear systems:

$$\begin{aligned} x_j^{(\alpha)} &= F_j(x_j, u_j) \\ y_j &= C_j x_j + D_j u_j \end{aligned} \tag{7.4}$$

where $1 \leq j \leq p$ denotes the $j-th$ system, $x_j \in \mathbb{R}^{n_j}$ is the state vector, $F_j(\cdot)$ is a nonlinear vector function, u_j is the input, y_j is the output and C_j, D_j are matrices of appropriate size.

We establish the following important result.

Lemma 7.1 *Consider the family of nonlinear systems (7.4). If the output is chosen as:*

$$y_j = \sum_{i=n-n_j+1}^{n} \gamma_i x_i + \sum_{k}^{m} \beta_k u_k,$$

where γ_i, β_k are differential quantities of u and their time finite derivatives, such that the first component of the coordinate transformation is given by $\bar{y}_j = I^\alpha y_j$, then the nonlinear system (7.4) is transformable to a MFGOCF if and only if is a family of PV systems.

Proof Let the set $\{\zeta_j, \zeta_j^{(\alpha)}, \ldots, \mathscr{D}^{[n_j-1]\alpha}\zeta_j, \}$, $1 \leq j \leq p$ with $\zeta_j = I^{(\alpha)}y_j = \bar{y}_j$, $\mathscr{D}^{[i-l]\alpha}\zeta_j = \mathscr{D}^{[i-l-1]\alpha}y_j$, $1 \leq i \leq \sum_{j=1}^{p} n_j$, where $n_j \geq 0$ is the minimum integer such that $\mathscr{D}^{[n_j-1]\alpha}y_j$ is dependent on $I^{(\alpha)}y_j$, y, $y_j^{(\alpha)}, \ldots, \mathscr{D}^{[n_j-l-1]\alpha}y_j$, u_j, ... Now, if we redefine $\xi_i = \zeta_j = I^{(\alpha)}y_j$, $\xi_i^{n_j} = \mathscr{D}^{[i-l]\alpha}\zeta_j = \mathscr{D}^{[i-l-1]\alpha}y_j$, $1 \leq i \leq \sum_{j=1}^{p} n_j$ that yields to:

$$\mathscr{D}^{\alpha}\xi_i^{n_1} = \xi_{i+1}^{n_1}, \quad 1 \leq i \leq n_1 - 1$$

$$\mathscr{D}^{\alpha}\xi_{n_1}^{n_1} = -\mathscr{L}_1(\xi_1^{n_1}, \ldots, \xi_{n_1}^{n_1}, u_1, u_1^{(\alpha)}, \ldots, \mathscr{D}^{[\gamma_1-1]\alpha}u_1) + \mathscr{D}^{\gamma_1\alpha}u_1$$

$$\mathscr{D}^{\alpha}\xi_i^{n_2} = \xi_{i+1}^{n_2}, \quad n_1 + 1 \leq i \leq n_1 + n_2 - 1$$

$$\mathscr{D}^{\alpha}\xi_{n_1+n_2}^{n_2} = -\mathscr{L}_2(\xi_{n_1+1}^{n_2}, \ldots, \xi_{n_1+n_2}^{n_2}, u_2, u_2^{(\alpha)}, \ldots, \mathscr{D}^{[\gamma_2-1]\alpha}u_2) + \mathscr{D}^{\gamma_2\alpha}u_2$$

$$\vdots$$

$$\mathscr{D}^{\alpha}\xi_i^{n_p} = \xi_{i+1}^{n_p}, \quad n_1 + \cdots + n_{p-1} + 1 \leq i \leq n_1 + \cdots + n_p - 1$$

$$\mathscr{D}^{\alpha}\xi_{n_1+\cdots+n_p}^{n_p} = -\mathscr{L}_p(\xi_{n_1+\cdots+n_{p-1}+1}^{n_p}, \ldots, \xi_{n_1+\cdots+n_p}^{n_p}, u_p, u_p^{(\alpha)}, \ldots, \mathscr{D}^{[\gamma_p-1]\alpha}u_p) + \mathscr{D}^{\gamma_p\alpha}u_p$$

$$\xi_i^{n_j} = I^{\alpha}y_j, \quad 1 \leq j \leq p, \quad l = 1 + \sum_{\hat{j}=1}^{j-1} n_{\hat{j}}.$$

$\square$

We discuss the problem of generalized synchronization for a class of fractional order systems so-called Fractional Liouvillian Systems, and within this class, we can find some fractional order chaotic systems. In this case, we consider the master–slave configuration. We define a family of master systems as:

$$\mathscr{D}^{(\alpha)}x_{m_\mu} = F_{m_\mu}(x_{m_\mu}, u_{m_\mu})$$
$$y_{m_\mu} = h_{m_\mu}(x_{m_\mu}, u_{m_\mu}) \tag{7.5}$$

and the family of slave systems is:

$$\mathscr{D}^{(\alpha)}x_{s_\nu} = F_{s_\nu}(x_{s_\nu}, u_{s_\nu})$$
$$y_{s_\nu} = h_{s_\nu}(x_{s_\nu}, u_{s_\nu}) \tag{7.6}$$

where $x_{s_\nu} = (x_{1,s_\nu}, \ldots, x_{n_{s_\nu},s_\nu}) \in \mathbb{R}^{n_{s_\nu}}$, $x_{m_\mu} = (x_{1,m_\mu}, \ldots, x_{n_{m_\mu},m_\mu}) \in \mathbb{R}^{n_{m_\mu}}$, $h_{s_\nu} : \mathbb{R}^{n_{s_\nu}} \to \mathbb{R}$, $h_{m_\mu} : \mathbb{R}^{n_{m_\mu}} \to \mathbb{R}$, $u_{m_\mu} = (u_{1,m_\mu}, \ldots, u_{\gamma_{m_\mu},m_\mu}) \in \mathbb{R}^{\gamma_{m_\mu}}$, $u_{s_\nu} = (u_{1,s_\nu}, \ldots, u_{\gamma_{s_\nu},s_\nu}) \in \mathbb{R}^{\gamma_{s_\nu}}$, $1 \leq \nu \leq p - 1$, $1 \leq \mu \leq p - \nu$, these conditions tell us that we can consider one or more slave systems associated with one master but we cannot have an slave with more than one master. One slave is associated with one master when the number of slaves is equal to the number of masters, on the other hand, it is possible to consider a case where the number of slaves is greater than the number of masters, this means that a master system interacts with more than one slave system. This configuration is depicted in Fig. 7.1, the circles or nodes represent the dynamical systems involved. Encircled nodes in Fig. 7.1b represent the same master system, dashed circles represent virtual master systems, these posses the same dynamics and initial conditions as the original master system (solid circles) associated with the corresponding slave system, hence it can be represented as a single master node.

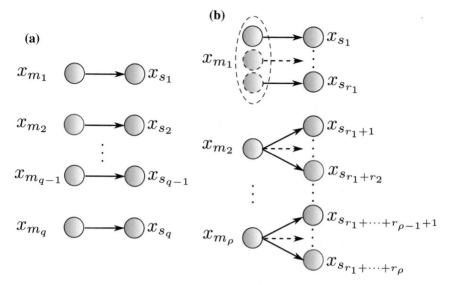

Fig. 7.1 Generalized Multi-synchronization configuration: **a** An equal number of slaves and masters, **b** More slaves than masters

From Definition 2.9, we can say that FGMS is achieved when $lim_{t\to\infty} \| H_{ms}(X_s) - X_m \| = 0$

Next theorem gives a methodology to synchronize multiple decoupled families of commensurate fractional order Liouvillian systems where systems need not to be identical (see Fig. 7.1). Moreover, the design of the dynamical controllers is given in proof. The idea behind it is to impose similar dynamics of the master systems to their corresponding family of slave systems.

Theorem 7.1 *Let a set of systems as (7.5) and (7.6) be transformable to a MFGOCF. Then,* $lim_{t\to\infty} \| \xi_m - \xi_s \| = 0$ *where* ξ_m *and* ξ_s *are the trajectories in the transformed space of the family of master and slave systems, respectively.*

Proof Without loss of generality, we consider that $u_{m_\mu} = 0$. The set of master systems has the following family of outputs:

$$y_{m_j} = \sum_{i=n-n_j+1}^{n} \gamma_i x_{i,m_j} = \xi_i^{n_{m_j}},$$

where γ_i, $(i = l)$ are differential quantities of u and their time finite derivatives, the family of outputs for slave systems is:

$$y_{s_j} = \sum_{i=n-n_j+1}^{n} \gamma_i x_{i,s_j} + \sum_k \beta_k u_{k,m_j} = \xi_i^{n_{s_j}},$$

where γ_i, $(i = l)$, β_k are differential quantities of u and their time finite derivatives. Then, we obtain:

$$\mathscr{D}^\alpha \xi_i^{n_{m_1}} = \xi_{i+1}^{n_{m_1}}, \quad 1 \leq i \leq n_{m_1} - 1$$

$$\mathscr{D}^\alpha \xi_{n_{m_1}}^{n_{m_1}} = -\mathscr{L}_{m_1}(\xi_1^{n_{m_1}}, \ldots, \xi_{n_{m_1}}^{n_{m_1}})$$

$$\mathscr{D}^\alpha \xi_i^{n_{m_2}} = \xi_{i+1}^{n_{m_2}}, \quad n_{m_1} + 1 \leq i \leq n_{m_1} + n_{m_2} - 1$$

$$\mathscr{D}^\alpha \xi_{n_{m_1}+n_{m_2}}^{n_{m_2}} = -\mathscr{L}_{m_2}(\xi_{n_{m_1}+1}^{n_{m_2}}, \ldots, \xi_{n_{m_1}+n_{m_2}}^{n_{m_2}})$$

$$\vdots$$

$$\mathscr{D}^\alpha \xi_i^{n_{m_p}} = \xi_{i+1}^{n_{m_p}}, \quad n_{m_1} + \cdots + n_{m_{p-1}} + 1 \leq i \leq n_{m_1} + \cdots + n_{m_p} - 1$$

$$\mathscr{D}^\alpha \xi_{n_{m_1}+\cdots+n_{m_p}}^{n_{m_p}} = -\mathscr{L}_{m_p}(\xi_{n_{m_1}+\cdots+n_{m_{p-1}}+1}^{n_{m_p}}, \ldots, \xi_{n_{m_1}+\cdots+n_{m_p}}^{n_{m_p}}),$$

$$(7.7)$$

in a compact form (7.7) can be expressed as:

$$\mathscr{D}^\alpha \xi_m = \mathscr{A}\xi_m - \Phi_m(\mathscr{L}_{m_1}, \ldots, \mathscr{L}_{m_p}) \tag{7.8}$$

Now, let us define the following extended system that represents the family of slave systems and a chain of fractional integrators given by a family of dynamical feedbacks is given as:

$$\mathscr{D}^\alpha \xi_i^{n_{s_1}} = \xi_{i+1}^{n_{s_1}}, \quad 1 \leq i \leq n_{s_1} - 1$$

$$\mathscr{D}^\alpha \xi_{n_{s_1}}^{n_{s_1}} = -\mathscr{L}_{s_1}(\xi_1^{n_{s_1}}, \ldots, \xi_{n_{s_1}}^{n_{s_1}}, u_{s_1}, u_{s_1}^{(\alpha)}, \ldots, \mathscr{D}^{[\gamma_{s_1}-1]\alpha}u_{s_1}) + \mathscr{D}^{\gamma_{s_1}\alpha}u_{s_1}$$

$$\mathscr{D}^\alpha u_i^{n_{s_1}} = u_{i+1}^{n_{s_1}}, \quad 1 \leq i \leq \gamma_{s_1} - 1$$

$$\mathscr{D}^\alpha u_{\gamma_{s_1}}^{n_{s_1}} = -\mathscr{L}_{m_1}(\xi_1^{n_{m_1}}, \ldots, \xi_{n_{m_1}}^{n_{m_1}}) + \mathscr{L}_{s_1}(\xi_1^{n_{s_1}}, \ldots, \xi_{n_{s_1}}^{n_{s_1}}, u_{s_1}, u_{s_1}^{(\alpha)}, \ldots, \mathscr{D}^{[\gamma_{s_1}-1]\alpha}u_{s_1})$$

$$\qquad + K_1(\xi^{n_{m_1}} - \xi^{n_{s_1}})$$

$$\mathscr{D}^\alpha \xi_i^{n_{s_2}} = \xi_{i+1}^{n_{s_2}}, \quad n_{s_1} + 1 \leq i \leq n_{s_1} + n_{s_2} - 1$$

$$\mathscr{D}^\alpha \xi_{n_{s_1}+n_{s_2}}^{n_{s_2}} = -\mathscr{L}_{s_2}(\xi_{n_{s_1}+1}^{n_{s_2}}, \ldots, \xi_{n_{s_1}+n_{s_2}}^{n_{s_2}}, u_{s_2}, u_{s_2}^{(\alpha)}, \ldots, \mathscr{D}^{[\gamma_{s_2}-1]\alpha}u_{s_2}) + \mathscr{D}^{\gamma_{s_2}\alpha}u_{s_2}$$

$$\mathscr{D}^\alpha u_i^{n_{s_2}} = u_{i+1}^{n_{s_2}}, \quad \gamma_{s_1} + 1 \leq i \leq \gamma_{s_1} + \gamma_{s_2} - 1$$

$$\mathscr{D}^\alpha u_{\gamma_{s_1}+\gamma_{s_2}}^{n_{s_2}} = -\mathscr{L}_{m_2}(\xi_{n_{m_1}+1}^{n_{m_2}}, \ldots, \xi_{n_{m_1}+n_{m_2}}^{n_{m_2}}) + \mathscr{L}_{s_2}(\xi_{n_{s_1}+1}^{n_{s_2}}, \ldots, \xi_{n_{s_1}+n_{s_2}}^{n_{s_2}}, u_{s_2}, u_{s_2}^{(\alpha)}, \ldots, \mathscr{D}^{[\gamma_{s_2}-1]\alpha}u_{s_2})$$

$$\qquad + K_2(\xi^{n_{m_2}} - \xi^{n_{s_2}})$$

$$\vdots$$

$$\mathscr{D}^\alpha \xi_i^{n_{s_p}} = \xi_{i+1}^{n_{s_p}}, \quad n_{s_1} + \cdots + n_{s_{p-1}} + 1 \leq i \leq n_{s_1} + \cdots + n_{s_p} - 1$$

$$\mathscr{D}^\alpha \xi_{n_{s_1}+\cdots+n_{s_p}}^{n_{s_p}} = -\mathscr{L}_{s_p}(\xi_{n_{s_1}+\cdots+n_{s_{p-1}}+1}^{n_{s_p}}, \ldots, \xi_{n_{s_1}+\cdots+n_{s_p}}^{n_{s_p}}, u_{s_p}, u_{s_p}^{(\alpha)}, \ldots, \mathscr{D}^{[\gamma_{s_p}-1]\alpha}u_{s_p}) + \mathscr{D}^{\gamma_{s_p}\alpha}u_{s_p}$$

$$\mathscr{D}^\alpha u_i^{n_{s_p}} = u_{i+1}^{n_{s_p}}, \quad \gamma_{s_1} + \cdots + \gamma_{s_{p-1}} + 1 \leq i \leq \gamma_{s_1} + \cdots + \gamma_{s_p} - 1$$

$$\mathscr{D}^\alpha u_{\gamma_{s_1}+\cdots+\gamma_{s_p}}^{n_{s_p}} = -\mathscr{L}_{m_p}(\xi^{n_{m_p}}) + \mathscr{L}_{s_p}(\xi^{n_{s_p}}, u_{s_p}, u_{s_p}^{(\alpha)}, \ldots, \mathscr{D}^{[\gamma_{s_p}-1]\alpha}u_{s_p}) + K_p(\xi^{n_{m_p}} - \xi^{n_{s_p}})$$

$$(7.9)$$

Rewriting (7.9) in a compact form, we have:

$$\mathscr{D}^\alpha \xi_s = \mathscr{A}\xi_s - \Phi_s(\mathscr{L}_{s_1},\ldots,\mathscr{L}_{s_p}) + \bar{\mathscr{U}}(\mathscr{D}^{\gamma_{s_1}\alpha} u_{s_1},\ldots,\mathscr{D}^{\gamma_{s_p}\alpha} u_{s_p})$$
$$\mathscr{D}^\alpha \mathscr{U} = \mathscr{M}\mathscr{U} + \bar{\mathscr{U}}$$
$$\bar{\mathscr{U}} = \mathscr{K}(\xi_m - \xi_s) - \Phi_m(\mathscr{L}_{m_1},\ldots,\mathscr{L}_{m_p}) + \Phi_s(\mathscr{L}_{s_1},\ldots,\mathscr{L}_{s_p})$$

where

$$\mathscr{U} = \begin{bmatrix} u_{n_{s_1}} \\ \vdots \\ u_{n_{s_p}} \end{bmatrix}, \quad u_{n_{s_j}} = \begin{bmatrix} u_1^{n_{s_j}} \\ u_2^{n_{s_j}} \\ \vdots \\ u_{\sum_{j=1}^{j}\gamma_{s_j}}^{n_{s_j}} \end{bmatrix}.$$

Assume the control signals as $u_1^{n_{s_j}} = u_{s_j}, u_2^{n_{s_j}} = \mathscr{D}^\alpha u_{s_j}, \ldots, u_{\alpha_{s_j}}^{n_{s_j}} = \mathscr{D}^{[\gamma_{s_j}-1]\alpha} u_{s_j}$,
$\xi^{n_{s_j}} = [\xi_{n_{s_1}+\cdots+n_{s_{j-1}}+1}^{n_{s_j}}, \ldots, \xi_{n_{s_1}+\cdots+n_{s_j}}^{n_{s_j}}]^T, \xi^{n_{m_j}} = [\xi_{n_{m_1}+\cdots+n_{m_{j-1}}+1}^{n_{m_j}}, \ldots, \xi_{n_{m_1}+\cdots+n_{m_j}}^{n_{m_j}}]^T$
and $K_j = [k_{1,j}, \ldots, k_{n_j,j}]$. Matrices $\mathscr{M}$ and $\mathscr{K}$ are defined as follows:

$$\mathscr{M} = \begin{bmatrix} \bar{\mathscr{M}}_1 & & 0 \\ & \ddots & \\ 0 & & \bar{\mathscr{M}}_p \end{bmatrix}, \quad \bar{\mathscr{M}}_j = \begin{bmatrix} 0 & 1 & 0 & 0 & \cdots & 0 \\ 0 & 0 & 1 & 0 & \cdots & 0 \\ \vdots & \vdots & \vdots & \ddots & \cdots & 0 \\ 0 & 0 & 0 & 0 & 1 & 0 \\ 0 & 0 & 0 & 0 & 0 & 1 \\ 0 & 0 & 0 & 0 & 0 & 0 \end{bmatrix},$$

$$\mathscr{K} = \begin{bmatrix} \bar{\mathscr{K}}_1 & & 0 \\ & \ddots & \\ 0 & & \bar{\mathscr{K}}_p \end{bmatrix}, \quad \bar{\mathscr{K}}_j = \begin{bmatrix} 0 & 0 & 0 & 0 & \cdots & 0 \\ 0 & 0 & 0 & 0 & \cdots & 0 \\ \vdots & \vdots & \vdots & \vdots & & \vdots \\ 0 & 0 & 0 & 0 & \cdots & 0 \\ 0 & 0 & 0 & 0 & \cdots & 0 \\ k_{1,j} & k_{2,j} & k_{3,j} & k_{4,j} & \cdots & k_{n_j,j} \end{bmatrix}.$$

Finally, we consider the error of synchronization $e_\xi = \xi_m - \xi_s$ that has a dynamics given by:

$$\mathscr{D}^\alpha e_\xi = \mathscr{A}\xi_m - \Phi_m(\mathscr{L}_{m_1},\ldots,\mathscr{L}_{m_p}) - \mathscr{A}\xi_s + \Phi_s(\mathscr{L}_{s_1},\ldots,\mathscr{L}_{s_p}) - \bar{\mathscr{U}}$$
$$\mathscr{D}^\alpha \mathscr{U} = \mathscr{M}\mathscr{U} + \bar{\mathscr{U}}$$
$$\bar{\mathscr{U}} = \mathscr{K}(\xi_m - \xi_s) - \Phi_m(\mathscr{L}_{m_1},\ldots,\mathscr{L}_{m_p}) + \Phi_s(\mathscr{L}_{s_1},\ldots,\mathscr{L}_{s_p})$$

and after some algebraic manipulations, we have that:

$$\mathscr{D}^{(\alpha)} e_\xi = (\mathscr{A} - \mathscr{K}) e_\xi \tag{7.10}$$

from the Theorem 2.3 (see Chap. 2), the system (7.10) is asymptotically stable if all eigenvalues of matrix $\mathscr{A} - \mathscr{K} = diag(\bar{\mathscr{A}}_1, \ldots, \bar{\mathscr{A}}_p)$ with control gains $(k_{1,j}, k_{2,j}, \ldots, k_{n_j,j})$ are chosen such that:

$$|arg(\lambda_i(\bar{\mathscr{A}}_j))| > \frac{\pi}{2} > \alpha\frac{\pi}{2}$$

where

$$\bar{\mathscr{A}}_j = \begin{bmatrix} 0 & 1 & 0 & 0 & \cdots & 0 \\ 0 & 0 & 1 & 0 & \cdots & 0 \\ \vdots & \vdots & \vdots & \ddots & \cdots & 0 \\ 0 & 0 & 0 & 0 & 1 & 0 \\ 0 & 0 & 0 & 0 & 0 & 1 \\ -k_{1,j} & -k_{2,j} & -k_{3,j} & -k_{4,j} & \cdots & -k_{n_j,j} \end{bmatrix}$$

$\square$

Corollary 7.1 *A family of Liouvillian fractional order system class that is a family of PV systems is in a state of FGMS.*

Proof The proof is trivial and it is omitted (transitivity). $\square$

7.3 Extension of Results to Complex Interaction Between Slave Systems

In this section, the previous results are extended to a more complex interaction of slave systems. This is the case when time-invariant interaction exists between slave systems of the same group, and each slave system have complete access to their corresponding master's system dynamics. Consider the master system's dynamics available for all slave systems of the same group as is described in Theorem 7.1, as well as consider a number of slave systems associated with a single master system. This is the case when the number of master systems is less than the number of slave systems. Let us define q as the number of groups of slave systems in the network. The Theorem 7.1 can be naturally extended to the case of complex interaction between slave members of the same family as is shown in Fig. 7.2.

Each interaction in Fig. 7.2 is represented by a dashed semicircle, it can be unidirectional or bidirectional depending on whether a slave system knows the state of its neighbors. Note that there is no link between slave systems from different groups. Hence, a complex system is considered as a population of interacting systems as described above. The interaction between "r_ρ" slave systems of the same group is modeled by the graph $G_{r_\rho} = (V_{r_\rho}, E_{r_\rho}, A_{r_\rho})$ with $1 \leq \rho \leq q$, let $V_{r_\rho} = \{1, \ldots, r_\rho\}$ a set of nodes, $E_{r_\rho} \subseteq V_{r_\rho} \times V_{r_\rho}$ a set of edges, and $A_{r_\rho} = [a_{ij}] \in \mathbb{R}^{r_\rho \times r_\rho}$ an adjacency matrix with nonnegative adjacency elements a_{ij}^ρ is defined by:

Fig. 7.2 Slave interactions in complex network

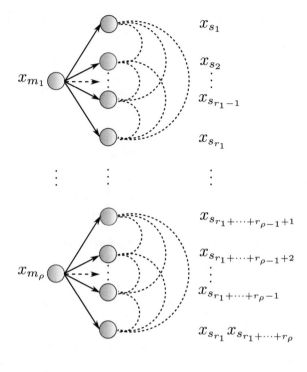

$$a_{ij}^{\rho} = \begin{cases} 1 & \text{if } (j,i) \in E_{r_{\rho}}, \\ 0 & \text{elsewhere.} \end{cases}$$

Let $L_{r_{\rho}} = [l_{ij}] \in \mathbb{R}^{r_{\rho} \times r_{\rho}}$ be the nonsymmetrical graph Laplacian matrix induced by the information flow $G_{r_{\rho}}$ that is defined as:

$$l_{ij}^{\rho} = \begin{cases} \sum_{k=1, k \neq i}^{r+1} a_{ik} & \text{if } i = j, \\ -a_{ij} & \text{elsewhere.} \end{cases}$$

Now, consider again the chain of integrators of the dynamical fractional order control law in Theorem 7.2, next we will add the following diffusive coupling terms Θ_{σ} in the dynamical control law for the σ-th slave system.

$$\Theta_{\sigma} = \sum_{j=1}^{r_{\rho}} a_{\varsigma(\sigma,\rho),j}^{\rho} \left[\kappa_1 \left(\xi_{\ell_n(\sigma)+1}^{ns_{\sigma}} - \xi_{\ell(\rho,j)+1}^{ns_{\vartheta(\rho,j)}} \right) + \dots + \kappa_{nm_{\rho}} \left(\xi_{\ell_n(\sigma)+ns_{\sigma}}^{ns_{\sigma}} - \xi_{\ell(\rho,j)+ns_{\vartheta(\rho,j)}}^{ns_{\vartheta(\rho,j)}} \right) \right]$$

with $\ell_n(\sigma) = \sum_{l=1}^{\sigma} n_{s_{l-1}}$, $\ell(\rho, j) = \sum_{l=1}^{\vartheta(\rho,j)} n_{s_{l-1}}$, $\vartheta(\rho, j) = j + r_{\rho-1}$, $\varsigma(\sigma, \rho) = \sigma - \sum_{l=1}^{\rho} r_{l-1}$, $1 \leq \varsigma(\sigma, \rho) \leq r_{\rho}$, $n_{s_0} = r_0 = 0$ and $\kappa_1, \dots, \kappa_{nm_{\rho}} > 0$ weighting scalar gains.

The interaction G_{r_ρ} is considered time invariant and fixed for all time. Incorporating this interaction in the network allows to obtain the following important result.

Theorem 7.2 *Let a set of systems as (7.5) and (7.6) with complex interaction Θ_σ for the σ-th slave system be transformable to a MFGOCF, then $\lim_{t\to\infty} \|\xi_m - \xi_s\| = 0$ where ξ_m and ξ_s are the trajectories in the transformed space of the family of master and slave systems, respectively.*

Proof Without loss of generality, we consider that $u_{m_\mu} = 0$. The set of master systems has the following family of outputs:

$$
y_{m_j} = \sum_{i=n-n_j+1}^{n} \gamma_i x_{i,m_j} = \xi_i^{n_{m_j}},
$$

where γ_i, $(i = l)$ are differential quantities of u and their time finite derivatives, the family of outputs for slave systems is:

$$
y_{s_j} = \sum_{i=n-n_j+1}^{n} \gamma_i x_{i,s_j} + \sum_k \beta_k u_{k,m_j} = \xi_i^{n_{s_j}},
$$

where γ_i, $(i = l)$, β_k are differential quantities of u and their time finite derivatives. Then, we obtain:

$$
\begin{aligned}
\mathscr{D}^\alpha \xi_i^{n_{m_1}} &= \xi_{i+1}^{n_{m_1}}, \quad 1 \le i \le n_{m_1} - 1 \\
\mathscr{D}^\alpha \xi_{n_{m_1}}^{n_{m_1}} &= -\mathscr{L}_{m_1}(\xi_1^{n_{m_1}}, \ldots, \xi_{n_{m_1}}^{n_{m_1}}) \\
\mathscr{D}^\alpha \xi_i^{n_{m_2}} &= \xi_{i+1}^{n_{m_2}}, \quad n_{m_1} + 1 \le i \le n_{m_1} + n_{m_2} - 1 \\
\mathscr{D}^\alpha \xi_{n_{m_1}+n_{m_2}}^{n_{m_2}} &= -\mathscr{L}_{m_2}(\xi_{n_{m_1}+1}^{n_{m_2}}, \ldots, \xi_{n_{m_1}+n_{m_2}}^{n_{m_2}}) \\
&\vdots \\
\mathscr{D}^\alpha \xi_i^{n_{m_\rho}} &= \xi_{i+1}^{n_{m_\rho}}, \quad n_{m_1} + \cdots + n_{m_{\rho-1}} + 1 \le i \le n_{m_1} + \cdots + n_{m_\rho} - 1 \\
\mathscr{D}^\alpha \xi_{n_{m_1}+\cdots+n_{m_\rho}}^{n_{m_\rho}} &= -\mathscr{L}_{m_\rho}(\xi_{n_{m_1}+\cdots+n_{m_{\rho-1}}+1}^{n_{m_\rho}}, \ldots, \xi_{n_{m_1}+\cdots+n_{m_\rho}}^{n_{m_\rho}}),
\end{aligned}
$$

$$(7.11)$$

with $1 \le \rho \le q \le p$. Now, let us define the following extended system which represents a group of r_ρ slave systems and a chain of fractional integrators given by a family of dynamical feedbacks:

$$\mathscr{D}^\alpha \xi_i^{n_{s_1}} = \xi_{i+1}^{n_{s_1}}, \quad 1 \le i \le n_{s_1} - 1$$

$$\mathscr{D}^\alpha \xi_{n_{s_1}}^{n_{s_1}} = -\mathscr{L}_{s_1}(\xi_1^{n_{s_1}}, \dots, \xi_{n_{s_1}}^{n_{s_1}}, u_{s_1}, u_{s_1}^{(\alpha)}, \dots, \mathscr{D}^{[\gamma_{s_1}-1]\alpha} u_{s_1}) + \mathscr{D}^{\gamma_{s_1}\alpha} u_{s_1}$$

$$\mathscr{D}^\alpha u_i^{n_{s_1}} = u_{i+1}^{n_{s_1}}, \quad 1 \le i \le \gamma_{s_1} - 1$$

$$\mathscr{D}^\alpha u_{\gamma_{s_1}}^{n_{s_1}} = -\mathscr{L}_{m_\rho}(\xi_{n_{m_1}+\dots+n_{m_{\rho-1}}+1}^{n_{m_\rho}}, \dots, \xi_{n_{m_1}+\dots+n_{m_\rho}}^{n_{m_\rho}}) +$$

$$+ \mathscr{L}_{s_1}(\xi_1^{n_{s_1}}, \dots, \xi_{n_{s_1}}^{n_{s_1}}, u_{s_1}, u_{s_1}^{(\alpha)}, \dots, \mathscr{D}^{[\gamma_{s_1}-1]\alpha} u_{s_1}) + K_1(\xi^{n_{m_\rho}} - \xi^{n_{s_1}}) +$$

$$+ \sum_{j=1}^{r_\rho} a_{1j}^\rho \left[\kappa_1 \left(\xi_1^{n_{s_1}} - \xi_{\ell(\rho,j)+1}^{n_{s_{\vartheta(\rho,j)}}} \right) + \dots + \kappa_{n_{m_\rho}} \left(\xi_{n_{s_1}}^{n_{s_1}} - \xi_{\ell(\rho,j)+n_{s_{\vartheta(\rho,j)}}}^{n_{s_{\vartheta(\rho,j)}}} \right) \right]$$

$$\mathscr{D}^\alpha \xi_i^{n_{s_2}} = \xi_{i+1}^{n_{s_2}}, \quad n_{s_1}+1 \le i \le n_{s_1}+n_{s_2} - 1$$

$$\mathscr{D}^\alpha \xi_{n_{s_1}+n_{s_2}}^{n_{s_2}} = -\mathscr{L}_{s_2}(\xi_{n_{s_1}+1}^{n_{s_2}}, \dots, \xi_{n_{s_1}+n_{s_2}}^{n_{s_2}}, u_{s_2}, u_{s_2}^{(\alpha)}, \dots, \mathscr{D}^{[\gamma_{s_2}-1]\alpha} u_{s_2}) + \mathscr{D}^{\gamma_{s_2}\alpha} u_{s_2}$$

$$\mathscr{D}^\alpha u_i^{n_{s_2}} = u_{i+1}^{n_{s_2}}, \quad \gamma_{s_1}+1 \le i \le \gamma_{s_1}+\gamma_{s_2} - 1$$

$$\mathscr{D}^\alpha u_{\gamma_{s_1}+\gamma_{s_2}}^{n_{s_2}} = -\mathscr{L}_{m_\rho}(\xi_{n_{m_1}+\dots+n_{m_{\rho-1}}+1}^{n_{m_\rho}}, \dots, \xi_{n_{m_1}+\dots+n_{m_\rho}}^{n_{m_\rho}}) +$$

$$+ \mathscr{L}_{s_2}(\xi_{n_{s_1}+1}^{n_{s_2}}, \dots, \xi_{n_{s_1}+n_{s_2}}^{n_{s_2}}, u_{s_2}, u_{s_2}^{(\alpha)}, \dots, \mathscr{D}^{[\gamma_{s_2}-1]\alpha} u_{s_2}) + K_2(\xi^{n_{m_\rho}} - \xi^{n_{s_2}}) +$$

$$+ \sum_{j=1}^{r_\rho} a_{2j}^\rho \left[\kappa_1 \left(\xi_{n_{s_1}+1}^{n_{s_2}} - \xi_{\ell(\rho,j)+1}^{n_{s_{\vartheta(\rho,j)}}} \right) + \dots + \kappa_{n_{m_\rho}} \left(\xi_{n_{s_1}+n_{s_2}}^{n_{s_2}} - \xi_{\ell(\rho,j)+n_{s_{\vartheta(\rho,j)}}}^{n_{s_{\vartheta(\rho,j)}}} \right) \right]$$

$$\vdots$$

$$\mathscr{D}^\alpha \xi_i^{n_{s_{r_\rho}}} = \xi_{i+1}^{n_{s_{r_\rho}}}, \quad n_{s_1}+\dots+n_{s_{r_\rho}-1}+1 \le i \le n_{s_1}+\dots+n_{s_{r_\rho}} - 1$$

$$\mathscr{D}^\alpha \xi_{n_{s_1}+\dots+n_{s_{r_\rho}}}^{n_{s_{r_\rho}}} = -\mathscr{L}_{s_{r_\rho}}(\xi_{n_{s_1}+\dots+n_{s_{r_\rho}-1}+1}^{n_{s_{r_\rho}}}, \dots, \xi_{n_{s_1}+\dots+n_{s_{r_\rho}}}^{n_{s_{r_\rho}}}, u_{s_{r_\rho}}, u_{s_{r_\rho}}^{(\alpha)}, \dots, \mathscr{D}^{[\gamma_{s_{r_\rho}}-1]\alpha} u_{s_{r_\rho}}) + \mathscr{D}^{\gamma_{s_{r_\rho}}\alpha} u_{s_{r_\rho}}$$

$$\mathscr{D}^\alpha u_i^{n_{s_{r_\rho}}} = u_{i+1}^{n_{s_{r_\rho}}}, \quad \gamma_{s_1}+\dots+\gamma_{s_{r_\rho}-1}+1 \le i \le \gamma_{s_1}+\dots+\gamma_{s_{r_\rho}} - 1$$

$$\mathscr{D}^\alpha u_{\gamma_{s_1}+\dots+\gamma_{s_{r_\rho}}}^{n_{s_{r_\rho}}} = -\mathscr{L}_{m_\rho}(\xi_{n_{m_1}+\dots+n_{m_{\rho-1}}+1}^{n_{m_\rho}}, \dots, \xi_{n_{m_1}+\dots+n_{m_\rho}}^{n_{m_\rho}}) +$$

$$+ \mathscr{L}_{s_{r_\rho}}(\xi^{n_{s_{r_\rho}}}, u_{s_{r_\rho}}, u_{s_{r_\rho}}^{(\alpha)}, \dots, \mathscr{D}^{[\gamma_{s_{r_\rho}}-1]\alpha} u_{s_{r_\rho}}) + K_{r_\rho}(\xi^{n_{m_\rho}} - \xi^{n_{s_{r_\rho}}}) +$$

$$+ \sum_{j=1}^{r_\rho} a_{r_\rho j}^\rho \left[\kappa_1 \left(\xi_{n_{s_1}+\dots+n_{s_{r_\rho}-1}+1}^{n_{s_{r_\rho}}} - \xi_{\ell(\rho,j)+1}^{n_{s_{\vartheta(\rho,j)}}} \right) + \dots + \right.$$

$$\left. + \kappa_{n_{m_\rho}} \left(\xi_{n_{s_1}+\dots+n_{s_{r_\rho}}}^{n_{s_{r_\rho}}} - \xi_{\ell(\rho,j)+n_{s_{\vartheta(\rho,j)}}}^{n_{s_{\vartheta(\rho,j)}}} \right) \right]$$

in a general compact form, we have:

$$\mathscr{D}^\alpha \xi_i^{n_{s_\sigma}} = \xi_{i+1}^{n_{s_\sigma}}, \quad \ell_n(\sigma)+1 \le i \le \ell_n(\sigma)+n_{s_\sigma} - 1$$

$$\mathscr{D}^\alpha \xi_{\ell_n(\sigma)+n_{s_\sigma}}^{n_{s_\sigma}} = -\mathscr{L}_{s_\sigma}(\xi_{\ell_n(\sigma)+1}^{n_{s_\sigma}}, \dots, \xi_{\ell_n(\sigma)+n_{s_\sigma}}^{n_{s_\sigma}}, u_{s_\sigma}, u_{s_\sigma}^{(\alpha)}, \dots, \mathscr{D}^{[\gamma_{s_\sigma}-1]\alpha} u_{s_\sigma}) + \mathscr{D}^{\gamma_{s_\sigma}\alpha} u_{s_\sigma}$$

$$\mathscr{D}^\alpha u_i^{n_{s_\sigma}} = u_{i+1}^{n_{s_\sigma}}, \quad \ell_\gamma(\sigma)+1 \le i \le \ell_\gamma(\sigma)+\gamma_{s_\sigma} - 1$$

$$\mathscr{D}^\alpha u_{\ell_\gamma(\sigma)+\gamma_{s_\sigma}}^{n_{s_\sigma}} = -\mathscr{L}_{m_\rho}(\xi_{n_{m_1}+\dots+n_{m_{\rho-1}}+1}^{n_{m_\rho}}, \dots, \xi_{n_{m_1}+\dots+n_{m_\rho}}^{n_{m_\rho}}) +$$

$$+ \mathscr{L}_{s_\sigma}(\xi^{n_{s_\sigma}}, u_{s_\sigma}, u_{s_\sigma}^{(\alpha)}, \dots, \mathscr{D}^{[\gamma_{s_\sigma}-1]\alpha} u_{s_\sigma}) + K_\sigma(\xi^{n_{m_\rho}} - \xi^{n_{s_\sigma}}) +$$

$$+ \sum_{j=1}^{r_\rho} a_{\varsigma(\sigma,\rho),j}^\rho \left[\kappa_1 \left(\xi_{\ell_n(\sigma)+1}^{n_{s_\sigma}} - \xi_{\ell(\rho,j)+1}^{n_{s_{\vartheta(\rho,j)}}} \right) + \dots + \kappa_{n_{m_\rho}} \left(\xi_{\ell_n(\sigma)+n_{s_\sigma}}^{n_{s_\sigma}} - \xi_{\ell(\rho,j)+n_{s_{\vartheta(\rho,j)}}}^{n_{s_{\vartheta(\rho,j)}}} \right) \right]$$

$$(7.12)$$

with $\ell_n(\sigma) = \sum_{l=1}^\sigma n_{s_{l-1}}$, $\ell_\gamma(\sigma) = \sum_{l=1}^\sigma \gamma_{s_{l-1}}$, $\ell(\rho,j) = \sum_{l=1}^{\vartheta(\rho,j)} n_{s_{l-1}}$, $\vartheta(\rho,j) = j + r_{\rho-1}$, $\varsigma(\sigma,\rho) = \sigma - \sum_{l=1}^\rho r_{l-1}$, $1 \le \varsigma(\sigma,\rho) \le r_\rho$, $n_{s_0} = r_0 = 0$. Assume the control signals as $u_1^{n_{s_\sigma}} = u_{s_\sigma}, u_2^{n_{s_\sigma}} = \mathscr{D}^\alpha u_{s_\sigma}, \dots, u_{\alpha_{s_\sigma}}^{n_{s_\sigma}} = \mathscr{D}^{[\gamma_{s_\sigma}-1]\alpha} u_{s_\sigma}$, $\xi^{n_{s_\sigma}} = [\xi_{n_{s_1}+\dots+n_{s_{\sigma-1}}+1}^{n_{s_\sigma}}, \dots, \xi_{n_{s_1}+\dots+n_{s_\sigma}}^{n_{s_\sigma}}]^T, \xi^{n_{m_\rho}} = [\xi_{n_{m_1}+\dots+n_{m_{\rho-1}}+1}^{n_{m_\rho}}, \dots, \xi_{n_{m_1}+\dots+n_{m_\rho}}^{n_{m_\rho}}]^T, K_\sigma = [k_{1,\sigma}, \dots, k_{n_\sigma,\sigma}]$ and scalars $\kappa_1, \dots, \kappa_{n_{m_\rho}} > 0$.

Define the synchronization error for each slave's state as:

$$e_{\xi_i}^{n_{s_\sigma}} := \xi_{\hat{\ell}_n(\rho)+i-\ell_n(\sigma)}^{n_{m_\rho}} - \xi_i^{n_{s_\sigma}} \tag{7.13}$$

for $\ell_n(\sigma)+1 \le i \le \ell_n(\sigma)+n_{s_\sigma}$, with $\hat{\ell}_n = \sum_{l=1}^{\rho} n_{m_{l-1}}$, $n_{m_0} = 0$. From equations (7.11) and (7.12), it is clear that the fractional order dynamics of synchronization error (7.13) are given by:

$$\mathscr{D}^\alpha e_{\xi_i}^{n_{s_\sigma}} = e_{\xi_{i+1}} \quad \ell_n(\sigma)+1 \le i \le \ell(\sigma)+n_{s_\sigma}-1$$

$$\mathscr{D}^\alpha e_{\xi_{\ell_n(\sigma)+n_{s_\sigma}}}^{n_{s_\sigma}} = -K_\sigma e_\xi^{n_{s_\sigma}} +$$

$$+ \sum_{j=1}^{r_\rho} a_{\varsigma(\sigma,\rho),j}^\rho \left[\kappa_1 \left(e_{\xi_{\ell_n(\sigma)+1}}^{n_{s_\sigma}} - e_{\xi(\rho,j)+1}^{n_{s_{\vartheta(\rho,j)}}} \right) + \ldots + \kappa_{n_{m_\rho}} \left(e_{\xi_{\ell_n(\sigma)+n_{s_\sigma}}}^{n_{s_\sigma}} - e_{\xi(\rho,j)+n_{s_{\vartheta(\rho,j)}}}^{n_{s_{\vartheta(\rho,j)}}} \right) \right]$$

$$\mathscr{D}^\alpha u_i^{n_{s_\sigma}} = u_{i+1}^{n_{s_\sigma}}, \quad \ell_\gamma(\sigma)+1 \le i \le \ell_\gamma(\sigma)+\gamma_{s_\sigma}-1 \tag{7.14}$$

$$\mathscr{D}^\alpha u_{\ell_\gamma(\sigma)+\gamma_{s_\sigma}}^{n_{s_\sigma}} = -\mathscr{L}_{m_\rho}(\xi_{n_{m_1}+\ldots+n_{m_{\rho-1}}+1}^{n_{m_\rho}}, \ldots, \xi_{n_{m_1}+\ldots+n_{m_\rho}}^{n_{m_\rho}}) +$$

$$+ \mathscr{L}_{s_\sigma}(\xi^{n_{s_\sigma}}, u_{s_\sigma}, u_{s_\sigma}^{(\alpha)}, \ldots, \mathscr{D}^{[\gamma_{s_\sigma}-1]\alpha} u_{s_\sigma}) + K_\sigma(\xi^{n_{m_\rho}} - \xi^{n_{s_\sigma}}) +$$

$$+ \sum_{j=1}^{r_\rho} a_{\varsigma(\sigma,\rho),j}^\rho \left[\kappa_1 \left(\xi_{\ell_n(\sigma)+1}^{n_{s_\sigma}} - \xi_{\ell(\rho,j)+1}^{n_{s_{\vartheta(\rho,j)}}} \right) + \ldots + \kappa_{n_{m_\rho}} \left(\xi_{\ell_n(\sigma)+n_{s_\sigma}}^{n_{s_\sigma}} - \xi_{\ell(\rho,j)+n_{s_{\vartheta(\rho,j)}}}^{n_{s_{\vartheta(\rho,j)}}} \right) \right]$$

with $e_\xi^{n_{s_\sigma}} = [e_{\xi_{\ell_n(\sigma)+1}}^{n_{s_\sigma}}, \ldots, e_{\xi_{\ell_n(\sigma)+n_{s_\sigma}}}^{n_{s_\sigma}}]^T$. Define matrix B_ρ as follows:

$$B_\rho = \begin{pmatrix} 0 & 0 & 0 & \ldots & 0 \\ 0 & 0 & 0 & \ldots & 0 \\ \vdots & \vdots & \vdots & & \vdots \\ \kappa_1 & \kappa_2 & \kappa_3 & \ldots & \kappa_{n_{m_\rho}} \end{pmatrix} \in \mathbb{R}^{n_{m_\rho} \times n_{m_\rho}}$$

and assume Laplacian matrix L_{r_ρ} then the closed-loop synchronization error dynamics for the group of r_ρ slave systems is given by ($\otimes$ Kronecker product, see Definition 2.13 Chap. 2):

$$\mathscr{D}^\alpha e_\xi^{r_\rho} = \underbrace{\left(-L_{r_\rho} \otimes B_\rho + diag(\bar{\mathscr{A}}_{\vartheta(\rho,1)}, \ldots, \bar{\mathscr{A}}_{\vartheta(\rho,r_\rho)}) \right)}_{\Xi_\rho} e_\xi^{r_\rho}$$

with $e_\xi^{r_\rho} = \left[e_\xi^{n_{s_{\vartheta(\rho,1)}}}, \ldots, e_\xi^{n_{s_{\vartheta(\rho,r_\rho)}}} \right]^T$ and for the whole network, that is, the q slave groups in the network, we have:

$$\mathscr{D}^\alpha e_\xi = (-\mathscr{L} + \Xi) e_\xi \tag{7.15}$$

where $\mathscr{L} = diag\left(L_{r_1} \otimes B_1, \ldots, L_{r_q} \otimes B_q \right)$ and $\Xi = diag\left(\Xi_1, \ldots, \Xi_q \right)$.

From the Theorem 2.3 (see Chap. 2), the system (7.15) is asymptotically stable if eigenvalues of matrix $(-\mathscr{L} + \Xi)$ with control gains $(k_{1,\sigma}, \ldots, k_{n_\sigma,\sigma})$ and scalars $\kappa_1, \ldots, \kappa_{n_{m_\rho}} > 0$ are chosen such that:

$$|arg(\lambda_i(\varXi_\rho))| > \frac{\pi}{2} > \alpha\frac{\pi}{2} \qquad \qquad \Box$$

Remark 7.1 Note that there is no restriction on the complex interaction between slave systems, that is, synchronization is satisfied as a graph G_{r_ρ} can be considered either directed or undirected, with any adjacency matrix A_{r_ρ} and not necessarily identical systems. Notice that Theorem 7.2 is recovered if $A_{r_\rho} = 0_{r_\rho \times r_\rho}$.

Remark 7.2 The case of complex interaction given in [11] is contained in Theorem 7.1. It is worth mentioning that we are using dynamical controllers to synchronize multiple groups of slave systems.

7.4 Some Numerical Examples

In this section, we consider the FGMS of three strictly different commensurate fractional order systems. We show four examples with Rössler, Arneodo and Chua–Hartley systems in different configurations. It will be shown that our methodology is not restricted to Liouvillian-type systems, this is the case given in the last two networks involving Rössler systems, which are not Liouvillian systems. The first example considers the case of GS. The second example considers the case of FGMS with one master system where CS and GS is achieved. The third example consists of the case of two families of slave systems without interaction. And, final example shows the interesting case of complex interaction between slave systems with two master systems.

7.4.1 Example 1

Let a master–slave configuration with Arneodo and Chua–Hartley systems as master and slave respectively. This is depicted in Fig. 7.3. The objective is to achieve the state of GS and shows the Liouvillian feature of these systems.

First, let the master system be:

$$
\begin{aligned}
\mathscr{D}^\alpha x_1^{m_1} &= x_2^{m_1} \\
\mathscr{D}^\alpha x_2^{m_1} &= x_3^{m_1} \\
\mathscr{D}^\alpha x_3^{m_1} &= -\beta_1 x_1^{m_1} - \beta_2 x_2^{m_1} - \beta_3 x_3^{m_1} + \beta_4 x_1^{m_1 3}
\end{aligned}
\qquad (7.16)
$$

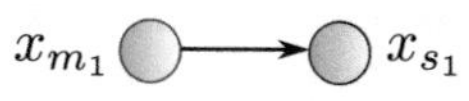

Fig. 7.3 GS configuration: Master system x_{m_1} and slave system x_{s_1}

assume $y_{m_1} = x_2^{m_1}$ as output, we obtain the states of system (7.16) as a function of the output:

$$\mathscr{D}^\alpha x_1^{m_1} = y_{m_1}$$
$$x_2^{m_1} = y_{m_1}$$
$$x_3 = y_{m_1}^{(\alpha)}$$

Hence, state $x_1^{m_1}$ satisfies FLAO condition. Now, note that $x_1^{m_1}$ can be written as a function of a fractional integral of $x_2^{m_1}$, that is to say:

$$x_1^{m_1} = I^\alpha y_{m_1}$$
$$x_2^{m_1} = y_{m_1}$$
$$x_3^{m_1} = y_{m_1}^{(\alpha)}$$

thus, system (7.16) is a fractional order Liouvillian system. On the other hand, let us verify the observability condition that slave system fulfils. Let the slave system be:

$$\mathscr{D}^\alpha x_1^{s_1} = \rho \left(x_2^{s_1} + \frac{x_1^{s_1} - 2x_1^{s_1 3}}{7} \right)$$
$$\mathscr{D}^\alpha x_2^{s_1} = x_1^{s_1} - x_2^{s_1} + x_3^{s_1} \qquad\qquad (7.17)$$
$$\mathscr{D}^\alpha x_3^{s_1} = -\beta x_2^{s_1}$$

Assume $y_{s_1} = x_2^{s_1}$ as output, we obtain the states of system (7.17) as a function of the output, this yields to the next expressions:

$$x_1^{s_1} = y_{s_1} + y_{s_1}^{(\alpha)} - x_3^{s_1}$$
$$x_2^{s_1} = y_{s_1}$$
$$\mathscr{D}^\alpha x_3^{s_1} = -\beta y_{s_1}$$

Hence, states $x_1^{s_1}$ and $x_3^{s_1}$ satisfy FLAO condition. Note that, we can choose $\bar{y}_{s_1} = I^\alpha y_{s_1} + u_1^{s_1}$ such that $x_1^{s_1}$ and $x_3^{s_1}$ can be obtained as a function of fractional integrals of $x_2^{s_1}$, that is to say:

$$x_1^{s_1} = y_{s_1} + y_{s_1}^{(\alpha)} + \beta I^\alpha y_{s_1} - u_1^{s_1}$$
$$x_2^{s_1} = y_{s_1}$$
$$x_3^{s_1} = -\beta I^\alpha y_{s_1} + u_1^{s_1}$$

system (7.17) is a fractional Liouvillian system. Now, assume systems (7.16) and (7.17). Consider the family of outputs for the master and slave systems, respectively, and are as follows:

$$\bar{y}_{m_1} = I^\alpha y_{m_1} = x_1^{m_1}$$

and

$$\bar{y}_{s_1} = I^\alpha y_{s_1} + u_1^{s_1} = -\frac{1}{\beta} x_3^{s_1} + u_1^{s_1}$$

From the family of outputs, next transformations are fulfilled for the master system:

$$\xi_m = \begin{pmatrix} \xi_1^{m_1} \\ \xi_2^{m_1} \\ \xi_3^{m_1} \end{pmatrix} = \begin{pmatrix} I^\alpha y_{m_1} \\ y_{m_1} \\ \mathscr{D}^\alpha y_{m_1} \end{pmatrix} = \begin{pmatrix} x_1^{m_1} \\ x_2^{m_1} \\ x_3^{m_1} \end{pmatrix} = \Phi_m(X_m)$$

with its inverse:

$$X_m = \begin{pmatrix} x_1^{m_1} \\ x_2^{m_1} \\ x_3^{m_1} \end{pmatrix} = \begin{pmatrix} \xi_1^{m_1} \\ \xi_2^{m_1} \\ \xi_3^{m_1} \end{pmatrix} = \Phi_m^{-1}(\xi_m)$$

For the slave system, we have:

$$\xi_s = \begin{pmatrix} \xi_1^{s_1} \\ \xi_2^{s_1} \\ \xi_3^{s_1} \end{pmatrix} = \begin{pmatrix} I^\alpha y_{s_1} + u_1^{s_1} \\ y_{s_1} + u_2^{s_1} \\ \mathscr{D}^\alpha y_{s_1} + u_3^{s_1} \end{pmatrix} = \begin{pmatrix} -\frac{1}{\beta} x_3^{s_1} + u_1^{s_1} \\ x_2^{s_1} + u_2^{s_2} \\ x_1^{s_1} - x_2^{s_1} + x_3^{s_1} + u_3^{s_3} \end{pmatrix} = \Phi_s(X_s)$$

with its inverse:

$$X_s = \begin{pmatrix} x_1^{s_1} \\ x_2^{s_1} \\ x_3^{s_1} \end{pmatrix} = \begin{pmatrix} \beta(\xi_1^{s_1} - u_1^{s_1}) + \xi_2^{s_1} - u_2^{s_1} + \xi_3^{s_1} - u_3^{s_1} \\ \xi_2^{s_1} - u_2^{s_1} \\ -\beta(\xi_1^{s_1} - u_1^{s_1}) \end{pmatrix} = \Phi_s^{-1}(\xi_s)$$

Then, the master in transformed coordinates is given by:

$$\mathscr{D}^\alpha \xi_1^{m_1} = \xi_2^{m_1}$$
$$\mathscr{D}^\alpha \xi_2^{m_1} = \xi_3^{m_1}$$
$$\mathscr{D}^\alpha \xi_3^{m_1} = -\mathscr{L}_{m_1}(\xi_1^{m_1}, \xi_2^{m_1}, \xi_3^{m_1})$$

with

$$\mathscr{L}_{m_1}(\cdot) = \beta_1 \left(\xi_1^{s_1} - u_1^{s_1} \right) + \beta_2 \left(\xi_2^{s_1} - u_2^{s_1} \right) + \beta_3 \left(\xi_3^{s_1} - u_3^{s_1} \right) - \beta_4 \left(\xi_1^{s_1} - u_1^{s_1} \right)^3$$

and the slave system in transformed coordinates is given by:

$$\mathscr{D}^\alpha \xi_1^{s_1} = \xi_2^{s_1}$$
$$\mathscr{D}^\alpha \xi_2^{s_1} = \xi_3^{s_1}$$
$$\mathscr{D}^\alpha \xi_3^{s_1} = -\mathscr{L}_{s_1} \left(\xi_1^{s_1}, \xi_2^{s_1}, \xi_3^{s_1}, u_1^{s_1}, u_2^{s_1}, u_3^{s_1} \right) + \bar{u}_{s_1}$$

$$\mathcal{D}^\alpha u_1^{s_1} = u_2^{s_1}$$
$$\mathcal{D}^\alpha u_2^{s_1} = u_3^{s_1}$$
$$\mathcal{D}^\alpha u_3^{s_1} = \bar{u}_{s_1}$$

with

$$\mathcal{L}_{s_1}(\cdot) = -\frac{\rho\beta}{7}\left(\xi_1^{s_1} - u_1^{s_1}\right) - \left(\frac{8\rho - 7\beta}{7}\right)\left(\xi_2^{s_1} - u_2^{s_1}\right) - \left(\frac{\rho - 7}{7}\right)\left(\xi_3^{s_1} - u_3^{s_1}\right)$$
$$+ \frac{2\rho}{7}\left(\beta(\xi_1^{s_1} - u_1^{s_1}) + (\xi_2^{s_1} - u_2^{s_1}) + (\xi_3^{s_1} - u_3^{s_1})\right)^3$$

The closed-loop dynamics for synchronization error $e_\xi = \xi_m - \xi_s$ is represented in the next augmented system:

$$\mathcal{D}^\alpha e_{\xi_1}^{s_1} = e_{\xi_2}^{s_1}$$
$$\mathcal{D}^\alpha e_{\xi_2}^{s_1} = e_{\xi_3}^{s_1}$$
$$\mathcal{D}^\alpha e_{\xi_3}^{s_1} = -\mathcal{L}_{m_1}(\xi_1^{m_1}, \xi_2^{m_1}, \xi_3^{m_1}) + \mathcal{L}_{s_1}\left(\xi_1^{s_1}, \xi_2^{s_1}, \xi_3^{s_1}, u_1^{s_1}, u_2^{s_1}, u_3^{s_1}\right) - \mathcal{D}^\alpha u_3^{s_1}$$
$$\mathcal{D}^\alpha u_1^{s_1} = u_2^{s_1}$$
$$\mathcal{D}^\alpha u_2^{s_1} = u_3^{s_1}$$
$$\mathcal{D}^\alpha u_3^{s_1} = -\mathcal{L}_{m_1}(\xi_1^{m_1}, \xi_2^{m_1}, \xi_3^{m_1}) + \mathcal{L}_{s_1}\left(\xi_1^{s_1}, \xi_2^{s_1}, \xi_3^{s_1}, u_1^{s_1}, u_2^{s_1}, u_3^{s_1}\right) + k_{s_1}e_\xi^{s_1}$$

we have that $\mathcal{D}^\alpha e_\xi = (\mathcal{A} - \mathcal{K})e_\xi$. Then, synchronization error converges asymptotically to zero if matrix $\mathcal{A} - \mathcal{K} = \bar{\mathcal{A}}_1$ is Hurwitz. Where

$$\bar{\mathcal{A}}_1 = \begin{bmatrix} 0 & 1 & 0 \\ 0 & 0 & 1 \\ -k_{1,1} & -k_{2,1} & -k_{3,1} \end{bmatrix}$$

This is given when $k_{1,1}, k_{1,3} > 0$ and $k_{1,2} > k_{1,1}/k_{1,3}$. Parameters for master and slave systems are $\rho = 12.75$, $\beta = 100/7$, $\beta_1 = -5.5$, $\beta_2 = 3.5$, $\beta_3 = 0.8$, $\beta_4 = -1$, commensurate fractional -order $\alpha = 0.92$ [17], initial conditions $x_{m_1}(0) = \begin{bmatrix} -0.20 & 0.35 & 0.20 \end{bmatrix}^T$, $x_{s_1}(0) = \begin{bmatrix} -0.58 & -0.01 & 0.30 \end{bmatrix}^T$ to obtain a chaotic behavior and $k_{s,1} = \begin{bmatrix} 10 & 10 & 10 \end{bmatrix}$.

In Figs. 7.4 and 7.5, GS is shown in original and transformed coordinates. This case of study is given for completeness the chapter. It illustrates how GS is given in master–slave configuration. Figures 7.4 and 7.5 are apparently the same due to master system is given in a canonical form (transformation $\Phi_m(\cdot)$ is equal to the identity), that makes the mapping

$$H_{ms}(X_s) = \Phi_m^{-1} \circ \Phi_s(X_s) = \begin{pmatrix} -\frac{1}{\beta}x_3^{s_1} + u_1^{s_1} \\ x_2^{s_1} + u_2^{s_2} \\ x_1^{s_1} - x_2^{s_1} + x_3^{s_1} + u_3^{s_3} \end{pmatrix} \quad (7.18)$$

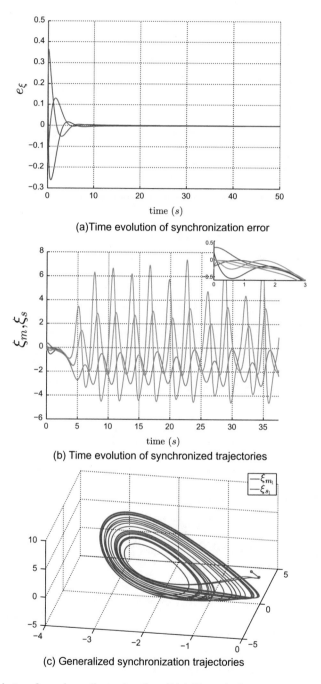

(a)Time evolution of synchronization error

(b) Time evolution of synchronized trajectories

(c) Generalized synchronization trajectories

Fig. 7.4 GS in transformed coordinates (see Sect. 7.4.1, Example 1)

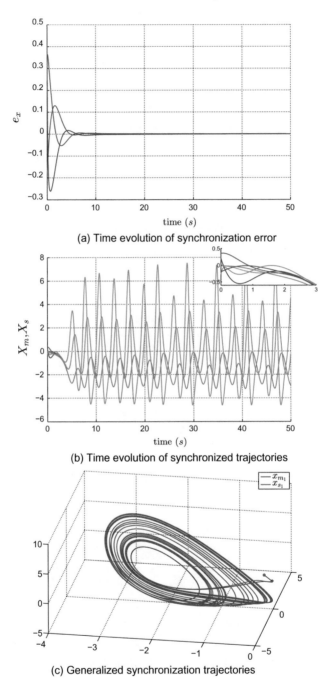

(a) Time evolution of synchronization error

(b) Time evolution of synchronized trajectories

(c) Generalized synchronization trajectories

Fig. 7.5 GS in original coordinates (see Sect. 7.4.1, Example 1)

to be easily obtained. The structure of the mapping is directly obtained from $\Phi_s(\cdot)$, which means that synchronized original and transformed coordinates are equal, therefore we obtain similar trajectories. Original trajectories for the slave system are obtained from mapping (7.18). Synchronization error asymptotically converges to the origin as seen in Figs. 7.4a and 7.5a. Finally, note that GS trajectories converge to the same chaotic attractor. Convergence can be verified from time evolution of the synchronized trajectories (see Figs. 7.4b, c and 7.5b, c).

7.4.2 Example 2

Consider two strictly different commensurate fractional order Liouvillian systems. The first one is a Chua–Hartley system as master, and let Arneodo and Chua–Hartley be as slave systems. This configuration is illustrated in Fig. 7.6. The objective is to achieve the state of CS and GS for these Liouvillian systems, hence the state of FGMS.

Consider the master system (Chua–Hartley) which is given by:

$$
\begin{aligned}
\mathscr{D}^\alpha x_1^{m_1} &= \rho\left(x_2^{m_1} + \frac{x_1^{m_1} - 2x_1^{m_1 3}}{7}\right) \\
\mathscr{D}^\alpha x_2^{m_1} &= x_1^{m_1} - x_2^{m_1} + x_3^{m_1} \\
\mathscr{D}^\alpha x_3^{m_1} &= -\beta x_2^{m_1}
\end{aligned}
\tag{7.19}
$$

Assume $y_{m_1} = x_2^{m_1}$ as output, we obtain the states of system (7.19) as a function of the output, this yields to the next expressions:

$$
\begin{aligned}
x_1^{m_1} &= y_{m_1} + y_{m_1}^{(\alpha)} - x_3^{m_1} \\
x_2^{m_1} &= y_{m_1} \\
\mathscr{D}^\alpha x_3^{m_1} &= -\beta y_{m_1}
\end{aligned}
$$

Hence, states $x_1^{m_1}$ and $x_3^{m_1}$ satisfy FLAO condition. On the other hand, we rewrite $x_1^{m_1}$ and $x_3^{m_1}$ as a function of fractional integrals of $x_2^{m_1}$, that is to say:

$$
\begin{aligned}
x_1^{m_1} &= y_{m_1} + y_{m_1}^{(\alpha)} + \beta I^\alpha y_{m_1} \\
x_2^{m_1} &= y_{m_1} \\
x_3^{m_1} &= -\beta I^\alpha y_{m_1}
\end{aligned}
$$

system (7.19) is a Fractional Liouvillian system.

Fig. 7.6 Configuration of
master system x_{m_1} and slave
systems x_{s_1}, x_{s_2}

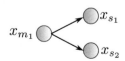

Now, consider the master system (7.19) and assume dynamics of the family of slave systems as:

$$\mathscr{D}^\alpha x_1^{s_1} = x_2^{s_1}$$
$$\mathscr{D}^\alpha x_2^{s_1} = x_3^{s_1}$$
$$\mathscr{D}^\alpha x_3^{s_1} = -\beta_1 x_1^{s_1} - \beta_2 x_2^{s_1} - \beta_3 x_3^{s_1} + \beta_4 x_1^{s_1\,3}$$
$$\mathscr{D}^\alpha x_4^{s_2} = \rho\left(x_5^{s_2} + \frac{x_4^{s_2} - 2x_4^{s_2\,3}}{7}\right)$$
$$\mathscr{D}^\alpha x_5^{s_2} = x_4^{s_2} - x_5^{s_2} + x_6^{s_2}$$
$$\mathscr{D}^\alpha x_6^{s_2} = -\beta x_5^{s_2}$$

Consider the family of outputs for the family of master and slave systems, respectively, are as follows:

$$\bar{y}_{m_1} = I^\alpha y_{m_1} = -\frac{1}{\beta} x_3^{m_1}$$

and

$$\bar{y}_{s_1} = I^\alpha y_{s_1} + u_1^{s_1} = x_1^{s_1} + u_1^{s_1}$$
$$\bar{y}_{s_2} = I^\alpha y_{s_2} + u_4^{s_2} = -\frac{1}{\beta} x_6^{s_2} + u_4^{s_2}$$

Then, the master system in transformed coordinates is given by:

$$\mathscr{D}^\alpha \xi_1^{m_1} = \xi_2^{m_1}$$
$$\mathscr{D}^\alpha \xi_2^{m_1} = \xi_3^{m_1}$$
$$\mathscr{D}^\alpha \xi_3^{m_1} = -\mathscr{L}_{m_1}(\xi_1^{m_1}, \xi_2^{m_1}, \xi_3^{m_1})$$

with

$$\mathscr{L}_{m_1}(\cdot) = -\frac{\rho\beta}{7}\xi_1^{m_1} - \left(\frac{8\rho - 7\beta}{7}\right)\xi_2^{m_1} - \left(\frac{\rho - 7}{7}\right)\xi_3^{m_1} + \frac{2\rho}{7}\left(\xi_1^{m_1} + \xi_2^{m_1} + \xi_3^{m_1}\right)^3$$

and the family of slave systems in transformed coordinates is given by:

$$\mathscr{D}^\alpha \xi_1^{s_1} = \xi_2^{s_1}$$
$$\mathscr{D}^\alpha \xi_2^{s_1} = \xi_3^{s_1}$$
$$\mathscr{D}^\alpha \xi_3^{s_1} = -\mathscr{L}_{s_1}\left(\xi_1^{s_1}, \xi_2^{s_1}, \xi_3^{s_1}, u_1^{s_1}, u_2^{s_1}, u_3^{s_1}\right) + \bar{u}_{s_1}$$
$$\mathscr{D}^\alpha u_1^{s_1} = u_2^{s_1}$$
$$\mathscr{D}^\alpha u_2^{s_1} = u_3^{s_1}$$

$$\mathscr{D}^{\alpha} u_3^{s_1} = \bar{u}_{s_1}$$
$$\mathscr{D}^{\alpha} \xi_4^{s_2} = \xi_5^{s_2}$$
$$\mathscr{D}^{\alpha} \xi_5^{s_2} = \xi_6^{s_2}$$
$$\mathscr{D}^{\alpha} \xi_6^{s_2} = -\mathscr{L}_{s_2} \left(\xi_4^{s_2}, \xi_5^{s_2}, \xi_6^{s_2}, u_4^{s_2}, u_5^{s_2}, u_6^{s_2} \right) + \bar{u}_{s_2}$$
$$\mathscr{D}^{\alpha} u_4^{s_2} = u_5^{s_2}$$
$$\mathscr{D}^{\alpha} u_5^{s_2} = u_6^{s_2}$$
$$\mathscr{D}^{\alpha} u_6^{s_2} = \bar{u}_{s_2}$$

with

$$\mathscr{L}_{s_1} (\cdot) = \beta_1 \left(\xi_1^{s_1} - u_1^{s_1} \right) + \beta_2 \left(\xi_2^{s_1} - u_2^{s_1} \right) + \beta_3 \left(\xi_3^{s_1} - u_3^{s_1} \right) - \beta_4 \left(\xi_1^{s_1} - u_1^{s_1} \right)^3$$

$$\mathscr{L}_{s_2} (\cdot) = -\frac{\rho\beta}{7} \left(\xi_4^{s_2} - u_4^{s_2} \right) - \left(\frac{8\rho - 7\beta}{7} \right) \left(\xi_5^{s_2} - u_5^{s_2} \right) - \left(\frac{\rho - 7}{7} \right) \left(\xi_6^{s_2} - u_6^{s_2} \right)$$

$$+ \frac{2\rho}{7} \left(\beta (\xi_4^{s_2} - u_4^{s_2}) + (\xi_5^{s_2} - u_5^{s_2}) + (\xi_6^{s_2} - u_6^{s_2}) \right)^3$$

Remark 7.3 We extend the dimension of master systems via virtual master systems that will have the same dynamics and initial conditions as the original master systems associated with the corresponding slave system (see Fig. 7.1a).

The closed-loop dynamics of synchronization error $e_\xi = \xi_m - \xi_s$ is given by the following augmented system:

$$\mathscr{D}^{\alpha} e_{\xi_1}^{s_1} = e_{\xi_2}^{s_1}$$
$$\mathscr{D}^{\alpha} e_{\xi_2}^{s_1} = e_{\xi_3}^{s_1}$$
$$\mathscr{D}^{\alpha} e_{\xi_3}^{s_1} = -\mathscr{L}_{m_1} (\xi_1^{m_1}, \xi_2^{m_1}, \xi_3^{m_1}) + \mathscr{L}_{s_1} \left(\xi_1^{s_1}, \xi_2^{s_1}, \xi_3^{s_1}, u_1^{s_1}, u_2^{s_1}, u_3^{s_1} \right) - \mathscr{D}^{\alpha} u_3^{s_1}$$
$$\mathscr{D}^{\alpha} u_1^{s_1} = u_2^{s_1}$$
$$\mathscr{D}^{\alpha} u_2^{s_1} = u_3^{s_1}$$
$$\mathscr{D}^{\alpha} u_3^{s_1} = -\mathscr{L}_{m_1} (\xi_1^{m_1}, \xi_2^{m_1}, \xi_3^{m_1}) + \mathscr{L}_{s_1} \left(\xi_1^{s_1}, \xi_2^{s_1}, \xi_3^{s_1}, u_1^{s_1}, u_2^{s_1}, u_3^{s_1} \right) + k_{s_1} e_{\xi}^{s_1}$$

$$\mathscr{D}^{\alpha} e_{\xi_4}^{s_2} = e_{\xi_5}^{s_2}$$
$$\mathscr{D}^{\alpha} e_{\xi_5}^{s_2} = e_{\xi_6}^{s_2}$$
$$\mathscr{D}^{\alpha} e_{\xi_6}^{s_2} = -\mathscr{L}_{m_1} (\xi_1^{m_1}, \xi_2^{m_1}, \xi_3^{m_1}) + \mathscr{L}_{s_2} \left(\xi_4^{s_2}, \xi_5^{s_2}, \xi_6^{s_2}, u_4^{s_2}, u_5^{s_2}, u_6^{s_2} \right) - \mathscr{D}^{\alpha} u_6^{s_2}$$
$$\mathscr{D}^{\alpha} u_4^{s_2} = u_5^{s_2}$$
$$\mathscr{D}^{\alpha} u_5^{s_2} = u_6^{s_2}$$
$$\mathscr{D}^{\alpha} u_6^{s_2} = -\mathscr{L}_{m_1} (\xi_1^{m_1}, \xi_2^{m_1}, \xi_3^{m_1}) + \mathscr{L}_{s_2} \left(\xi_4^{s_2}, \xi_5^{s_2}, \xi_6^{s_2}, u_4^{s_2}, u_5^{s_2}, u_6^{s_2} \right) + k_{s_2} e_{\xi}^{s_2}$$

After some algebraic manipulations, we have that $\mathcal{D}^{\alpha} e_{\xi} = (\mathcal{A} - \mathcal{K}) e_{\xi}$. Then, synchronization error converges asymptotically to zero if matrix $\mathcal{A} - \mathcal{K} = \text{diag}(\bar{\mathcal{A}}_1, \bar{\mathcal{A}}_2)$ is Hurwitz. Where

$$\bar{\mathcal{A}}_j = \begin{bmatrix} 0 & 1 & 0 \\ 0 & 0 & 1 \\ -k_{1,j} & -k_{2,j} & -k_{3,j} \end{bmatrix}, \quad 1 \leq j \leq 2.$$

Parameters are taken as in example 1 (Sect. 7.4.1). Let initial conditions be $x_{m_1}(0) = \begin{bmatrix} -0.50 & -0.07 & 0.65 \end{bmatrix}^T$, $x_{s_1}(0) = \begin{bmatrix} -0.20 & 0.35 & 0.20 \end{bmatrix}^T$, $x_{s_2}(0) = \begin{bmatrix} -0.58 & -0.01 & 0.30 \end{bmatrix}^T$ to ensure chaotic behavior and $k_{s,j} = \begin{bmatrix} 200 & 200 & 200 \end{bmatrix}$ for $1 \leq j \leq 2$.

Figure 7.7 illustrates transformed coordinates of master and slave systems that are in state of FGMS, Fig. 7.7a shows error synchronization convergence in transformed coordinates, and in Fig. 7.8 are shown the states of master and slave systems and synchronization error convergence in original coordinates with

$$H_{ms}(X_s) = \begin{pmatrix} \beta(x_1^{s_1} + u_1^{s_1}) + x_2^{s_1} + u_2^{s_1} + x_3^{s_1} + u_3^{s_1} \\ x_2^{s_1} + u_2^{s_1} \\ -\beta(x_1^{s_1} + u_1^{s_1}) \\ x_4^{s_2} + \beta u_4^{s_2} + u_5^{s_2} + u_6^{s_2} \\ x_5^{s_2} + u_5^{s_2} \\ -\beta\left(x_4^{s_2} - x_5^{s_2} + x_6^{s_2} + u_6^{s_2}\right) \end{pmatrix}$$

The effectiveness of our approach can be verified from the multi-sychronization trajectories. Note that the synchronization error converges asymptotically to the origin.

7.4.3 Example 3

Assume the configuration composed of two master systems given in Fig. 7.9. The objective is to synchronize the decoupled groups of slave systems with their associated master system.

The first group is considered as the system given in example above. Consider the second master as a Rössler system with Arneodo, Chua and Rössler systems as slaves. Let the second master system be:

$$\begin{aligned} \mathcal{D}^{\alpha} x_4^{m_2} &= -(x_5^{m_2} + x_6^{m_2}) \\ \mathcal{D}^{\alpha} x_5^{m_2} &= x_4^{m_2} + a x_5^{m_2} \\ \mathcal{D}^{\alpha} x_6^{m_2} &= b + x_6^{m_2}(x_4^{m_2} - c) \end{aligned} \qquad (7.20)$$

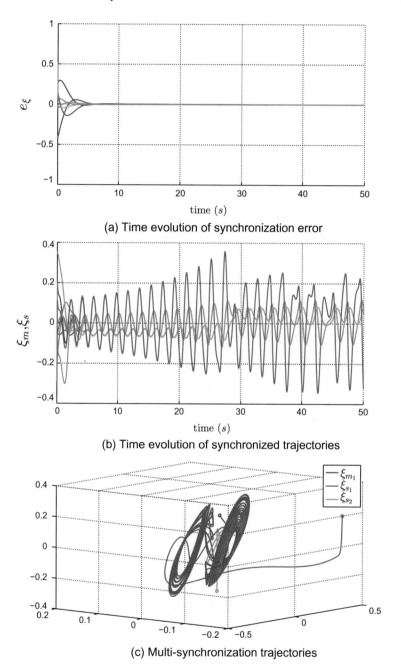

(a) Time evolution of synchronization error

(b) Time evolution of synchronized trajectories

(c) Multi-synchronization trajectories

Fig. 7.7   FGMS in transformed coordinates (see Sect. 7.4.2, Example 2)

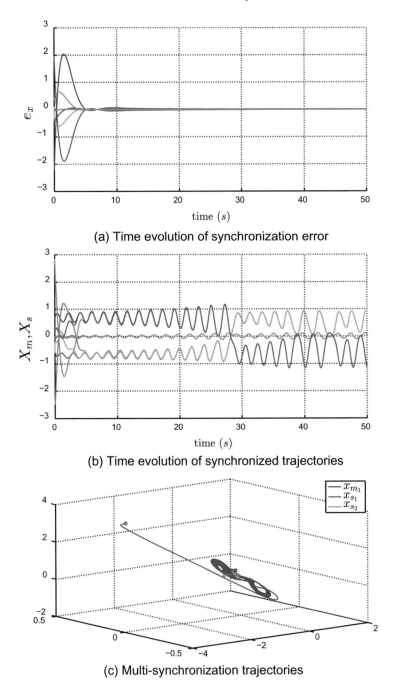

(a) Time evolution of synchronization error

(b) Time evolution of synchronized trajectories

(c) Multi-synchronization trajectories

Fig. 7.8 FGMS in original coordinates (see Sect. 7.4.2, Example 2)

Fig. 7.9 Configuration of master system x_{m_1} and slave systems x_{s_1}, x_{s_2}

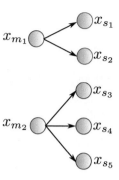

assume $y_{m_2} = x_5^{m_2}$ as output, we obtain the states of system (7.20) as a function of the output:

$$x_4^{m_2} = \mathscr{D}^{\alpha} y_{m_2} + a y_{m_2}$$
$$x_5^{m_2} = y_{m_2}$$
$$x_6^{m_2} = -\mathscr{D}^{2\alpha} y_{m_2} - \mathscr{D}^{\alpha} y_{m_2} + y_{m_2}$$

Hence, the states of system (7.20) satisfy FAO condition. Consider the family of outputs for the family of master and slave systems, respectively, are as follows:

$$\bar{y}_{m_1} = I^{\alpha} y_{m_1} = -\frac{1}{\beta} x_3^{m_1}$$
$$\bar{y}_{m_2} = y_{m_2} = x_5^{m_2}$$

and

$$\bar{y}_{s_1} = I^{\alpha} y_{s_1} + u_1^{s_1} = x_1^{s_1} + u_1^{s_1}$$
$$\bar{y}_{s_2} = I^{\alpha} y_{s_2} + u_4^{s_2} = -\frac{1}{\beta} x_6^{s_2} + u_1^{s_2}$$
$$\bar{y}_{s_3} = I^{\alpha} y_{s_3} + u_7^{s_3} = x_7^{s_3} + u_7^{s_3}$$
$$\bar{y}_{s_4} = I^{\alpha} y_{s_4} + u_{10}^{s_4} = -\frac{1}{\beta} x_{12}^{s_4} + u_{10}^{s_4}$$
$$\bar{y}_{s_5} = y_{s_5} + u_{13}^{s_5} = x_{14}^{s_5} + u_{13}^{s_5}$$

Then, the master in transformed coordinates is given by:

$$\mathscr{D}^{\alpha} \xi_1^{m_1} = \xi_2^{m_1}$$
$$\mathscr{D}^{\alpha} \xi_2^{m_1} = \xi_3^{m_1}$$
$$\mathscr{D}^{\alpha} \xi_3^{m_1} = -\mathscr{L}_{m_1}(\xi_1^{m_1}, \xi_2^{m_1}, \xi_3^{m_1})$$
$$\mathscr{D}^{\alpha} \xi_4^{m_2} = \xi_5^{m_2}$$
$$\mathscr{D}^{\alpha} \xi_5^{m_2} = \xi_6^{m_2}$$

$$\mathscr{D}^{\alpha}\xi_6^{m_2} = -\mathscr{L}_{m_2}(\xi_4^{m_2}, \xi_5^{m_2}, \xi_6^{m_2})$$

with

$$\mathscr{L}_{m_1}(\cdot) = -\frac{\rho\beta}{7}\xi_1^{m_1} - \left(\frac{8\rho - 7\beta}{7}\right)\xi_2^{m_1} - \left(\frac{\rho - 7}{7}\right)\xi_3^{m_1} + \frac{2\rho}{7}\left(\xi_1^{m_1} + \xi_2^{m_1} + \xi_3^{m_1}\right)^3$$

$$\mathscr{L}_{m_2}(\cdot) = -(ac-1)\xi_5^{m_2} + c\xi_4^{m_2} + (c-a)\xi_6^{m_2} - (\xi_4^{m_2} - a\xi_5^{m_2} + \xi_6^{m_2})(\xi_5^{m_2} - a\xi_4^{m_2}) + b$$

and the family of slave systems in transformed coordinates is given by:

$$\mathscr{D}^{\alpha}\xi_1^{s_1} = \xi_2^{s_1}$$
$$\mathscr{D}^{\alpha}\xi_2^{s_1} = \xi_3^{s_1}$$
$$\mathscr{D}^{\alpha}\xi_3^{s_1} = -\mathscr{L}_{s_1}\left(\xi_1^{s_1}, \xi_2^{s_1}, \xi_3^{s_1}, u_1^{s_1}, u_2^{s_1}, u_3^{s_1}\right) + \bar{u}_{s_1}$$
$$\mathscr{D}^{\alpha}u_1^{s_1} = u_2^{s_1}$$
$$\mathscr{D}^{\alpha}u_2^{s_1} = u_3^{s_1}$$
$$\mathscr{D}^{\alpha}u_3^{s_1} = \bar{u}_{s_1}$$
$$\mathscr{D}^{\alpha}\xi_4^{s_2} = \xi_5^{s_2}$$
$$\mathscr{D}^{\alpha}\xi_5^{s_2} = \xi_6^{s_2}$$
$$\mathscr{D}^{\alpha}\xi_6^{s_2} = -\mathscr{L}_{s_2}\left(\xi_4^{s_2}, \xi_5^{s_2}, \xi_6^{s_2}, u_4^{s_2}, u_5^{s_2}, u_6^{s_2}\right) + \bar{u}_{s_2}$$
$$\mathscr{D}^{\alpha}u_4^{s_2} = u_5^{s_2}$$
$$\mathscr{D}^{\alpha}u_5^{s_2} = u_6^{s_2}$$
$$\mathscr{D}^{\alpha}u_6^{s_2} = \bar{u}_{s_2}$$
$$\mathscr{D}^{\alpha}\xi_7^{s_3} = \xi_8^{s_3}$$
$$\mathscr{D}^{\alpha}\xi_8^{s_3} = \xi_9^{s_3}$$
$$\mathscr{D}^{\alpha}\xi_9^{s_3} = -\mathscr{L}_{s_3}\left(\xi_7^{s_3}, \xi_8^{s_3}, \xi_9^{s_3}, u_7^{s_3}, u_8^{s_3}, u_9^{s_3}\right) + \bar{u}_{s_3}$$
$$\mathscr{D}^{\alpha}u_7^{s_3} = u_8^{s_3}$$
$$\mathscr{D}^{\alpha}u_8^{s_3} = u_9^{s_3}$$
$$\mathscr{D}^{\alpha}u_9^{s_3} = \bar{u}_{s_3}$$

$$\mathscr{D}^{\alpha}\xi_{10}^{s_4} = \xi_{11}^{s_4}$$
$$\mathscr{D}^{\alpha}\xi_{11}^{s_4} = \xi_{12}^{s_4}$$
$$\mathscr{D}^{\alpha}\xi_{12}^{s_4} = -\mathscr{L}_{s_4}\left(\xi_{10}^{s_4}, \xi_{11}^{s_4}, \xi_{12}^{s_4}, u_{10}^{s_4}, u_{11}^{s_4}, u_{12}^{s_4}\right) + \bar{u}_{s_4}$$
$$\mathscr{D}^{\alpha}u_{10}^{s_4} = u_{11}^{s_4}$$
$$\mathscr{D}^{\alpha}u_{11}^{s_4} = u_{12}^{s_4}$$
$$\mathscr{D}^{\alpha}u_{12}^{s_4} = \bar{u}_{s_4}$$
$$\mathscr{D}^{\alpha}\xi_{13}^{s_5} = \xi_{14}^{s_5}$$
$$\mathscr{D}^{\alpha}\xi_{14}^{s_5} = \xi_{15}^{s_5}$$
$$\mathscr{D}^{\alpha}\xi_{15}^{s_5} = -\mathscr{L}_{s_5}\left(\xi_{13}^{s_5}, \xi_{14}^{s_5}, \xi_{15}^{s_5}, u_{13}^{s_5}, u_{14}^{s_5}, u_{15}^{s_5}\right) + \bar{u}_{s_5}$$

$$\mathscr{D}^{\alpha} u_{13}^{s_5} = u_{14}^{s_5}$$
$$\mathscr{D}^{\alpha} u_{14}^{s_5} = u_{15}^{s_5}$$
$$\mathscr{D}^{\alpha} u_{15}^{s_5} = \bar{u}_{s_5}$$

with

$$\mathscr{L}_{s_1}(\cdot) = \beta_1 \left(\xi_1^{s_1} - u_1^{s_1}\right) + \beta_2 \left(\xi_2^{s_1} - u_2^{s_1}\right) + \beta_3 \left(\xi_3^{s_1} - u_3^{s_1}\right) - \beta_4 \left(\xi_1^{s_1} - u_1^{s_1}\right)^3$$

$$\mathscr{L}_{s_2}(\cdot) = -\frac{\rho\beta}{7} \left(\xi_4^{s_2} - u_4^{s_2}\right) - \left(\frac{8\rho - 7\beta}{7}\right) \left(\xi_5^{s_2} - u_5^{s_2}\right) - \left(\frac{\rho - 7}{7}\right) \left(\xi_6^{s_2} - u_6^{s_2}\right)$$
$$+ \frac{2\rho}{7} \left(\beta(\xi_4^{s_2} - u_4^{s_2}) + (\xi_5^{s_2} - u_5^{s_2}) + (\xi_6^{s_2} - u_6^{s_2})\right)^3$$

$$\mathscr{L}_{s_3}(\cdot) = \beta_1 \left(\xi_7^{s_3} - u_7^{s_3}\right) + \beta_2 \left(\xi_8^{s_3} - u_8^{s_3}\right) + \beta_3 \left(\xi_9^{s_3} - u_9^{s_3}\right) - \beta_4 \left(\xi_7^{s_3} - u_7^{s_3}\right)^3$$

$$\mathscr{L}_{s_4}(\cdot) = -\frac{\rho\beta}{7} \left(\xi_{10}^{s_4} - u_{10}^{s_4}\right) - \left(\frac{8\rho - 7\beta}{7}\right) \left(\xi_{11}^{s_4} - u_{11}^{s_4}\right) - \left(\frac{\rho - 7}{7}\right) \left(\xi_{12}^{s_4} - u_{12}^{s_4}\right)$$
$$+ \frac{2\rho}{7} \left(\beta(\xi_{10}^{s_4} - u_{10}^{s_4}) + (\xi_{11}^{s_4} - u_{11}^{s_4}) + (\xi_{12}^{s_4} - u_{12}^{s_4})\right)^3$$

$$\mathscr{L}_{s_5}(\cdot) = -(ac - 1)(\xi_{14}^{s_5} - u_{14}^{s_5}) + c(\xi_{13}^{s_5} - u_{13}^{s_5}) + (c - a)(\xi_{15}^{s_5} - u_{15}^{s_5})$$
$$- \left(\xi_{13}^{s_5} - u_{13}^{s_5} - a(\xi_{14}^{s_5} - u_{14}^{s_5}) + \xi_{15}^{s_5} - u_{15}^{s_5}\right) \left(\xi_{14}^{s_5} - u_{14}^{s_5} - a(\xi_{13}^{s_5} - u_{13}^{s_5})\right) + b$$

The closed-loop dynamics of synchronization error $e_{\xi} = \xi_m - \xi_s$ is given by the following augmented system:

$$\mathscr{D}^{\alpha} e_{\xi_1}^{s_1} = e_{\xi_2}^{s_1}$$
$$\mathscr{D}^{\alpha} e_{\xi_2}^{s_1} = e_{\xi_3}^{s_1}$$
$$\mathscr{D}^{\alpha} e_{\xi_3}^{s_1} = -\mathscr{L}_{m_1}(\xi_1^{m_1}, \xi_2^{m_1}, \xi_3^{m_1}) + \mathscr{L}_{s_1}\left(\xi_1^{s_1}, \xi_2^{s_1}, \xi_3^{s_1}, u_1^{s_1}, u_2^{s_1}, u_3^{s_1}\right) - \mathscr{D}^{\alpha} u_3^{s_1}$$
$$\mathscr{D}^{\alpha} u_1^{s_1} = u_2^{s_1}$$
$$\mathscr{D}^{\alpha} u_2^{s_1} = u_3^{s_1}$$
$$\mathscr{D}^{\alpha} u_3^{s_1} = -\mathscr{L}_{m_1}(\xi_1^{m_1}, \xi_2^{m_1}, \xi_3^{m_1}) + \mathscr{L}_{s_1}\left(\xi_1^{s_1}, \xi_2^{s_1}, \xi_3^{s_1}, u_1^{s_1}, u_2^{s_1}, u_3^{s_1}\right) + k_{s_1} e_{\xi}^{s_1}$$
$$\mathscr{D}^{\alpha} e_{\xi_4}^{s_2} = e_{\xi_5}^{s_2}$$
$$\mathscr{D}^{\alpha} e_{\xi_5}^{s_2} = e_{\xi_6}^{s_2}$$
$$\mathscr{D}^{\alpha} e_{\xi_6}^{s_2} = -\mathscr{L}_{m_1}(\xi_1^{m_1}, \xi_2^{m_1}, \xi_3^{m_1}) + \mathscr{L}_{s_2}\left(\xi_4^{s_2}, \xi_5^{s_2}, \xi_6^{s_2}, u_4^{s_2}, u_5^{s_2}, u_6^{s_2}\right) - \mathscr{D}^{\alpha} u_6^{s_2}$$
$$\mathscr{D}^{\alpha} u_4^{s_2} = u_5^{s_2}$$
$$\mathscr{D}^{\alpha} u_5^{s_2} = u_6^{s_2}$$
$$\mathscr{D}^{\alpha} u_6^{s_2} = -\mathscr{L}_{m_1}(\xi_1^{m_1}, \xi_2^{m_1}, \xi_3^{m_1}) + \mathscr{L}_{s_2}\left(\xi_4^{s_2}, \xi_5^{s_2}, \xi_6^{s_2}, u_4^{s_2}, u_5^{s_2}, u_6^{s_2}\right) + k_{s_2} e_{\xi}^{s_2}$$

$$\mathscr{D}^{\alpha} e_{\xi_7}^{s_3} = e_{\xi_8}^{s_3}$$
$$\mathscr{D}^{\alpha} e_{\xi_8}^{s_3} = e_{\xi_9}^{s_3}$$
$$\mathscr{D}^{\alpha} e_{\xi_9}^{s_3} = -\mathscr{L}_{m_2}(\xi_4^{m_2}, \xi_5^{m_2}, \xi_6^{m_2}) + \mathscr{L}_{s_3}\left(\xi_7^{s_3}, \xi_8^{s_3}, \xi_9^{s_3}, u_7^{s_3}, u_8^{s_3}, u_9^{s_3}\right) - \mathscr{D}^{\alpha} u_9^{s_3}$$

$$\mathscr{D}^{\alpha} u_7^{s_3} = u_8^{s_3}$$

$$\mathscr{D}^{\alpha} u_8^{s_3} = u_9^{s_3}$$

$$\mathscr{D}^{\alpha} u_9^{s_3} = -\mathscr{L}_{m_2}(\xi_4^{m_2}, \xi_5^{m_2}, \xi_6^{m_2}) + \mathscr{L}_{s_3}\left(\xi_7^{s_3}, \xi_8^{s_3}, \xi_9^{s_3}, u_7^{s_3}, u_8^{s_3}, u_9^{s_3}\right) + k_{s_3} e_{\xi}^{s_3}$$

$$\mathscr{D}^{\alpha} e_{\xi_{10}}^{s_4} = e_{\xi_{11}}^{s_4}$$

$$\mathscr{D}^{\alpha} e_{\xi_{11}}^{s_4} = e_{\xi_{12}}^{s_4}$$

$$\mathscr{D}^{\alpha} e_{\xi_{12}}^{s_4} = -\mathscr{L}_{m_2}(\xi_4^{m_2}, \xi_5^{m_2}, \xi_6^{m_2}) + \mathscr{L}_{s_4}\left(\xi_{10}^{s_4}, \xi_{11}^{s_4}, \xi_{12}^{s_4}, u_{10}^{s_4}, u_{11}^{s_4}, u_{12}^{s_4}\right) - \mathscr{D}^{\alpha} u_{12}^{s_4}$$

$$\mathscr{D}^{\alpha} u_{10}^{s_4} = u_{11}^{s_4}$$

$$\mathscr{D}^{\alpha} u_{11}^{s_4} = u_{12}^{s_4}$$

$$\mathscr{D}^{\alpha} u_{12}^{s_4} = -\mathscr{L}_{m_2}(\xi_4^{m_2}, \xi_5^{m_2}, \xi_6^{m_2}) + \mathscr{L}_{s_4}\left(\xi_{10}^{s_4}, \xi_{11}^{s_4}, \xi_{12}^{s_4}, u_{10}^{s_4}, u_{11}^{s_4}, u_{12}^{s_4}\right) + k_{s_4} e_{\xi}^{s_4}$$

$$\mathscr{D}^{\alpha} e_{\xi_{13}}^{s_5} = e_{\xi_{14}}^{s_5}$$

$$\mathscr{D}^{\alpha} e_{\xi_{14}}^{s_5} = e_{\xi_{15}}^{s_5}$$

$$\mathscr{D}^{\alpha} e_{\xi_{15}}^{s_5} = -\mathscr{L}_{m_2}(\xi_4^{m_2}, \xi_5^{m_2}, \xi_6^{m_2}) + \mathscr{L}_{s_5}\left(\xi_{13}^{s_5}, \xi_{14}^{s_5}, \xi_{15}^{s_4}, u_{13}^{s_5}, u_{14}^{s_5}, u_{15}^{s_5}\right) - \mathscr{D}^{\alpha} u_{15}^{s_5}$$

$$\mathscr{D}^{\alpha} u_{13}^{s_5} = u_{14}^{s_5}$$

$$\mathscr{D}^{\alpha} u_{14}^{s_5} = u_{15}^{s_5}$$

$$\mathscr{D}^{\alpha} u_{15}^{s_5} = -\mathscr{L}_{m_2}(\xi_4^{m_2}, \xi_5^{m_2}, \xi_6^{m_2}) + \mathscr{L}_{s_5}\left(\xi_{13}^{s_5}, \xi_{14}^{s_5}, \xi_{15}^{s_5}, u_{13}^{s_5}, u_{14}^{s_5}, u_{15}^{s_5}\right) + k_{s_5} e_{\xi}^{s_5}$$

After some algebraic manipulations, we have that $\mathscr{D}^{\alpha} e_{\xi} = (\mathscr{A} - \mathscr{K}) e_{\xi}$. Then, synchronization error converges asymptotically to zero if matrix $\mathscr{A} - \mathscr{K} = \mathrm{diag}\left(\bar{\mathscr{A}}_1, \bar{\mathscr{A}}_2, \bar{\mathscr{A}}_3, \bar{\mathscr{A}}_4, \bar{\mathscr{A}}_5\right)$ is Hurwitz. Where

$$\bar{\mathscr{A}}_j = \begin{bmatrix} 0 & 1 & 0 \\ 0 & 0 & 1 \\ -k_{1,j} & -k_{2,j} & -k_{3,j} \end{bmatrix}, \quad 1 \leq j \leq 5.$$

Parameters are taken as in example 1 with $a = 0.5$, $b = 0.2$ and $c = 10$ [17]. Let initial conditions be $x_{m_1}(0) = \begin{bmatrix} -0.50 & -0.07 & 0.65 \end{bmatrix}^T$, $x_{m_2}(0) = \begin{bmatrix} 0.50 & 1.5 & 0.1 \end{bmatrix}^T$, $x_{s_1}(0) = \begin{bmatrix} -0.20 & 0.35 & 0.20 \end{bmatrix}^T$, $x_{s_2}(0) = \begin{bmatrix} -0.58 & -0.01 & 0.30 \end{bmatrix}^T$, $x_{s_3}(0) = \begin{bmatrix} 2 & -0.1 & -2 \end{bmatrix}^T$, $x_{s_4}(0) = \begin{bmatrix} -0.71 & -0.1 & 0.45 \end{bmatrix}^T$, $x_{s_5}(0) = \begin{bmatrix} 1 & 2.5 & -1 \end{bmatrix}^T$ to ensure chaotic behavior and $k_{s,j} = \begin{bmatrix} 10 & 20 & 10 \end{bmatrix}$ for $1 \leq j \leq 5$.

Figure 7.10 illustrates synchronization error convergence to the origin in transformed and original coordinates, respectively, FGMS is shown in Figs. 7.11 and 7.12 in transformed coordinates. And, taking the mapping (7.21) FGMS is shown in original coordinates in Figs. 7.13 and 7.14. The effectiveness of our approach can be verified from the multi-synchronization trajectories. Here, multi-synchronization is shown for two decoupled groups. Note that the synchronization error converges asymptotically to the origin. In case of the first group, convergence to the trajectories of first master system is more slowly than Example 2 (see Figs. 7.7, 7.8, 7.11 and 7.13 respectively). This is due gains $k_{s,1}, k_{s,2}$ are smaller than gains in Example 2 (Sect. 7.4.2). It is worth mentioning that our methodology can be applied not only to Liouvillian systems, this is the case of any system that fulfils FAO condition, i.e., Rössler system.

$$
H_{ms}(X_s) = \begin{pmatrix}
\beta(x_1^{s_1} + u_1^{s_1}) + x_2^{s_1} + u_2^{s_1} + x_3^{s_1} + u_3^{s_1} \\
x_2^{s_1} + u_2^{s_1} \\
-\beta(x_1^{s_1} + u_1^{s_1}) \\
x_4^{s_2} + \beta u_4^{s_2} + u_5^{s_2} + u_6^{s_2} \\
x_5^{s_2} + u_5^{s_2} \\
-\beta\left(x_4^{s_2} - x_5^{s_2} + x_6^{s_2} + u_6^{s_2}\right) \\
u_8^{s_3} + x_8^{s_3} - a(x_7^{s_3} - u_7^{s_3}) \\
x_7^{s_3} + u_7^{s_3} \\
-(x_7^{s_3} - u_7^{s_3}) + a(x_8^{s_3} - u_8^{s_3}) - (x_9^{s_3} - u_9^{s_3}) \\
x_{11}^{s_4} + \frac{a}{\beta}x_{12}^{s_4} + u_{11}^{s_4} - au_{10}^{s_4} \\
-\frac{1}{\beta}x_{12}^{s_4} + u_{10} \\
-x_{10}^{s_4} + (a+1)x_{11}^{s_4} + \left(\frac{1}{\beta} - 1\right)x_{12}^{s_4} - u_{10}^{s_4} + au_{11}^{s_4} - u_{12}^{s_4} \\
x_{13}^{s_5} - au_{13}^{s_5} + u_{14}^{s_5} \\
x_{14}^{s_5} + u_{14}^{s_5} \\
x_{15}^{s_5} - u_{13}^{s_5} + au_{14}^{s_5} - u_{15}^{s_5}
\end{pmatrix} \tag{7.21}
$$

7.4.4 Example 4

Assume the following configuration which is composed of two master systems with complex interaction between slave systems:

The objective is to look at the state of FGMS in presence of interplay between slave systems of the same group. Consider all systems dynamics as in last example. Note that first interaction between slave systems is bidirectional (undirected graph G_{r_1}) and the second one is unidirectional (directed graph G_{r_2}) as depicted in Fig. 7.15. The associated graphs are given by:

$$
G_{r_1} = \{V_{r_1}, E_{r_1}, A_{r_1}\}, \quad G_{r_2} = \{V_{r_2}, E_{r_2}, A_{r_2}\}
$$

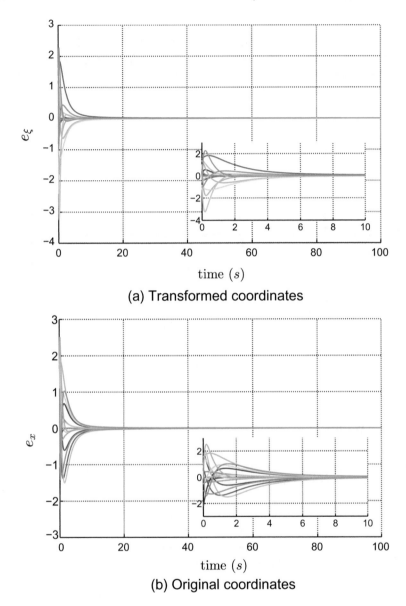

(a) Transformed coordinates

(b) Original coordinates

Fig. 7.10 Time evolution of synchronization error (see Sect. 7.4.3, Example 3)

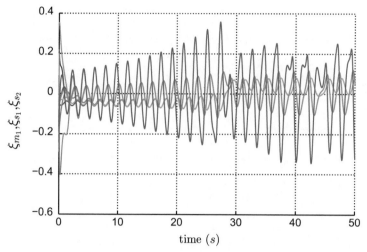

(a) Time evolution of synchronized trajectories

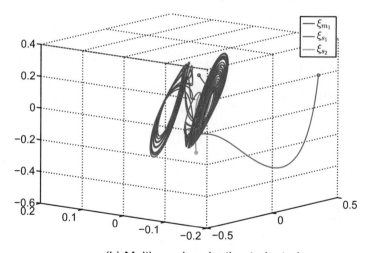

(b) Multi-synchronization trajectories

Fig. 7.11 FGMS in transformed coordinates (see Sect. 7.4.3, Example 3)

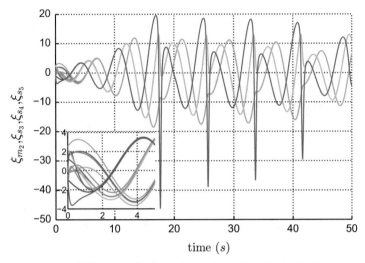

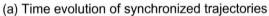

(a) Time evolution of synchronized trajectories

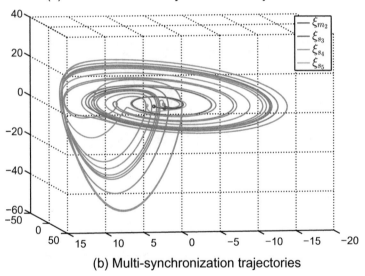

(b) Multi-synchronization trajectories

Fig. 7.12 FGMS in transformed coordinates (see Sect. 7.4.3, Example 3)

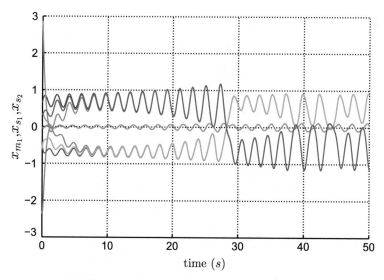

(a) Time evolution of synchronized trajectories

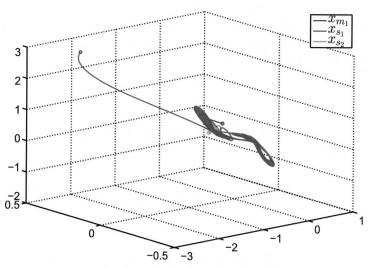

(b) Multi-synchronization trajectories

Fig. 7.13 FGMS in original coordinates (see Sect. 7.4.3, Example 3)

where

$$V_{r_1} = \{1, 2\}$$
$$E_{r_1} = \{(1, 2), (2, 1)\}$$
$$V_{r_2} = \{1, 2, 3\}$$
$$E_{r_2} = \{(1, 2), (2, 3), (3, 1)\}$$

and

$$A_{r_1} = \begin{pmatrix} 0 & 1 \\ 1 & 0 \end{pmatrix}, \qquad A_{r_2} = \begin{pmatrix} 0 & 1 & 0 \\ 0 & 0 & 1 \\ 1 & 0 & 0 \end{pmatrix}$$

respectively. It is easy to see that

$$L_{r_1} = \begin{pmatrix} 1 & -1 \\ -1 & 1 \end{pmatrix}, \qquad L_{r_2} = \begin{pmatrix} 1 & -1 & 0 \\ 0 & 1 & -1 \\ -1 & 0 & 1 \end{pmatrix}.$$

with $r_1 = 2$ and $r_2 = 3$, note that L_{r_1} is a symmetric matrix due to G_{r_1} is an undirected graph. The closed-loop dynamics of synchronization error $e_\xi = \xi_m - \xi_s$ is given by the following augmented system:

$$\mathscr{D}^\alpha e_{\xi_1}^{s_1} = e_{\xi_2}^{s_1}$$
$$\mathscr{D}^\alpha e_{\xi_2}^{s_1} = e_{\xi_3}^{s_1}$$
$$\mathscr{D}^\alpha e_{\xi_3}^{s_1} = -\mathscr{L}_{m_1}(\xi_1^{m_1}, \xi_2^{m_1}, \xi_3^{m_1}) + \mathscr{L}_{s_1}\left(\xi_1^{s_1}, \xi_2^{s_1}, \xi_3^{s_1}, u_1^{s_1}, u_2^{s_1}, u_3^{s_1}\right) - \mathscr{D}^\alpha u_3^{s_1}$$
$$\mathscr{D}^\alpha u_1^{s_1} = u_2^{s_1}$$
$$\mathscr{D}^\alpha u_2^{s_1} = u_3^{s_1}$$
$$\mathscr{D}^\alpha u_3^{s_1} = -\mathscr{L}_{m_1}(\xi_1^{m_1}, \xi_2^{m_1}, \xi_3^{m_1}) + \mathscr{L}_{s_1}\left(\xi_1^{s_1}, \xi_2^{s_1}, \xi_3^{s_1}, u_1^{s_1}, u_2^{s_1}, u_3^{s_1}\right) + k_{s_1} e_\xi^{s_1} + \Theta_1$$
$$\mathscr{D}^\alpha e_{\xi_4}^{s_2} = e_{\xi_5}^{s_2}$$
$$\mathscr{D}^\alpha e_{\xi_5}^{s_2} = e_{\xi_6}^{s_2}$$
$$\mathscr{D}^\alpha e_{\xi_6}^{s_2} = -\mathscr{L}_{m_1}(\xi_1^{m_1}, \xi_2^{m_1}, \xi_3^{m_1}) + \mathscr{L}_{s_2}\left(\xi_4^{s_2}, \xi_5^{s_2}, \xi_6^{s_2}, u_4^{s_2}, u_5^{s_2}, u_6^{s_2}\right) - \mathscr{D}^\alpha u_6^{s_2}$$
$$\mathscr{D}^\alpha u_4^{s_2} = u_5^{s_2}$$
$$\mathscr{D}^\alpha u_5^{s_2} = u_6^{s_2}$$
$$\mathscr{D}^\alpha u_6^{s_2} = -\mathscr{L}_{m_1}(\xi_1^{m_1}, \xi_2^{m_1}, \xi_3^{m_1}) + \mathscr{L}_{s_2}\left(\xi_4^{s_2}, \xi_5^{s_2}, \xi_6^{s_2}, u_4^{s_2}, u_5^{s_2}, u_6^{s_2}\right) + k_{s_2} e_\xi^{s_2} + \Theta_2$$
$$\mathscr{D}^\alpha e_{\xi_7}^{s_3} = e_{\xi_8}^{s_3}$$
$$\mathscr{D}^\alpha e_{\xi_8}^{s_3} = e_{\xi_9}^{s_3}$$
$$\mathscr{D}^\alpha e_{\xi_9}^{s_3} = -\mathscr{L}_{m_2}(\xi_4^{m_2}, \xi_5^{m_2}, \xi_6^{m_2}) + \mathscr{L}_{s_3}\left(\xi_7^{s_3}, \xi_8^{s_3}, \xi_9^{s_3}, u_7^{s_3}, u_8^{s_3}, u_9^{s_3}\right) - \mathscr{D}^\alpha u_9^{s_3}$$
$$\mathscr{D}^\alpha u_7^{s_3} = u_8^{s_3}$$

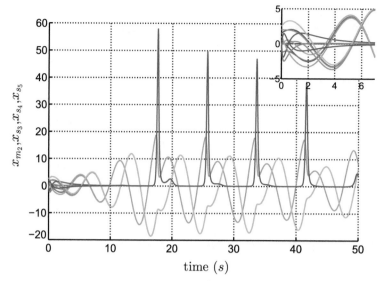

(a) Time evolution of synchronized trajectories

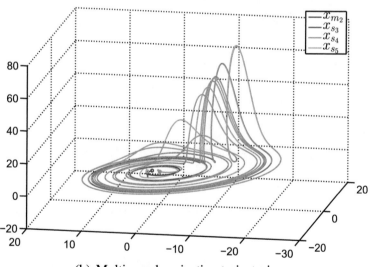

(b) Multi-synchronization trajectories

Fig. 7.14 FGMS in orignal coordinates (see Sect. 7.4.3, Example 3)

Fig. 7.15 Configuration of master system x_{m_1} and slave systems x_{s_1}, x_{s_2}

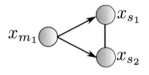

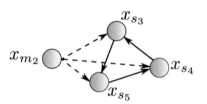

$$\mathscr{D}^\alpha u_8^{s_3} = u_9^{s_3}$$
$$\mathscr{D}^\alpha u_9^{s_3} = -\mathscr{L}_{m_2}(\xi_4^{m_2}, \xi_5^{m_2}, \xi_6^{m_2}) + \mathscr{L}_{s_3}\left(\xi_7^{s_3}, \xi_8^{s_3}, \xi_9^{s_3}, u_7^{s_3}, u_8^{s_3}, u_9^{s_3}\right) + k_{s_3} e_\xi^{s_3} + \Theta_3$$

$$\mathscr{D}^\alpha e_{\xi_{10}}^{s_4} = e_{\xi_{11}}^{s_4}$$
$$\mathscr{D}^\alpha e_{\xi_{11}}^{s_4} = e_{\xi_{12}}^{s_4}$$
$$\mathscr{D}^\alpha e_{\xi_{12}}^{s_4} = -\mathscr{L}_{m_2}(\xi_4^{m_2}, \xi_5^{m_2}, \xi_6^{m_2}) + \mathscr{L}_{s_4}\left(\xi_{10}^{s_4}, \xi_{11}^{s_4}, \xi_{12}^{s_4}, u_{10}^{s_4}, u_{11}^{s_4}, u_{12}^{s_4}\right) - \mathscr{D}^\alpha u_{12}^{s_4}$$
$$\mathscr{D}^\alpha u_{10}^{s_4} = u_{11}^{s_4}$$
$$\mathscr{D}^\alpha u_{11}^{s_4} = u_{12}^{s_4}$$
$$\mathscr{D}^\alpha u_{12}^{s_4} = -\mathscr{L}_{m_2}(\xi_4^{m_2}, \xi_5^{m_2}, \xi_6^{m_2}) + \mathscr{L}_{s_4}\left(\xi_{10}^{s_4}, \xi_{11}^{s_4}, \xi_{12}^{s_4}, u_{10}^{s_4}, u_{11}^{s_4}, u_{12}^{s_4}\right) + k_{s_4} e_\xi^{s_4} + \Theta_4$$
$$\mathscr{D}^\alpha e_{\xi_{13}}^{s_5} = e_{\xi_{14}}^{s_5}$$
$$\mathscr{D}^\alpha e_{\xi_{14}}^{s_5} = e_{\xi_{15}}^{s_5}$$
$$\mathscr{D}^\alpha e_{\xi_{15}}^{s_5} = -\mathscr{L}_{m_2}(\xi_4^{m_2}, \xi_5^{m_2}, \xi_6^{m_2}) + \mathscr{L}_{s_5}\left(\xi_{13}^{s_5}, \xi_{14}^{s_5}, \xi_{15}^{s_4}, u_{13}^{s_5}, u_{14}^{s_5}, u_{15}^{s_5}\right) - \mathscr{D}^\alpha u_{15}^{s_5}$$
$$\mathscr{D}^\alpha u_{13}^{s_5} = u_{14}^{s_5}$$
$$\mathscr{D}^\alpha u_{14}^{s_5} = u_{15}^{s_5}$$
$$\mathscr{D}^\alpha u_{15}^{s_5} = -\mathscr{L}_{m_2}(\xi_4^{m_2}, \xi_5^{m_2}, \xi_6^{m_2}) + \mathscr{L}_{s_5}\left(\xi_{13}^{s_5}, \xi_{14}^{s_5}, \xi_{15}^{s_5}, u_{13}^{s_5}, u_{14}^{s_5}, u_{15}^{s_5}\right) + k_{s_5} e_\xi^{s_5} + \Theta_5$$

where

$$\Theta_1 = \kappa_1\left(\xi_1^{s_1} - \xi_4^{s_2}\right) + \kappa_2\left(\xi_2^{s_1} - \xi_5^{s_2}\right) + \kappa_3\left(\xi_3^{s_1} - \xi_6^{s_2}\right)$$
$$\Theta_2 = -\Theta_1$$
$$\Theta_3 = \kappa_1\left(\xi_7^{s_3} - \xi_{10}^{s_4}\right) + \kappa_2\left(\xi_8^{s_3} - \xi_{11}^{s_4}\right) + \kappa_3\left(\xi_9^{s_3} - \xi_{12}^{s_4}\right)$$
$$\Theta_4 = \kappa_1\left(\xi_{10}^{s_4} - \xi_{13}^{s_5}\right) + \kappa_2\left(\xi_{11}^{s_4} - \xi_{14}^{s_5}\right) + \kappa_3\left(\xi_{12}^{s_4} - \xi_{15}^{s_5}\right)$$
$$\Theta_5 = \kappa_1\left(\xi_{13}^{s_5} - \xi_7^{s_3}\right) + \kappa_2\left(\xi_{14}^{s_5} - \xi_8^{s_3}\right) + \kappa_3\left(\xi_{15}^{s_5} - \xi_9^{s_3}\right)$$

After some algebraic manipulations, we have that $\mathscr{D}^\alpha e_\xi = (-\mathfrak{L} + \varXi)\, e_\xi$. Then, synchronization error converges asymptotically to zero if matrix $(-\mathfrak{L} + \varXi)$ is Hurwitz.

Where

$$\Xi = \text{diag}\,(\Xi_1, \Xi_2), \qquad \mathfrak{L} = \text{diag}\left(L_{r_1} \otimes B_1, L_{r_2} \otimes B_2\right),$$

$$\Xi_1 = -L_{r_1} \otimes B_1 + \text{diag}\left(\bar{\mathscr{A}}_1, \bar{\mathscr{A}}_2\right), \quad \Xi_2 = -L_{r_2} \otimes B_1 + \text{diag}\left(\bar{\mathscr{A}}_3, \bar{\mathscr{A}}_4, \bar{\mathscr{A}}_5\right),$$

$$B_1 = B_2 = \begin{pmatrix} 0 & 0 & 0 \\ 0 & 0 & 0 \\ \kappa_1 & \kappa_2 & \kappa_3 \end{pmatrix}, \quad \bar{\mathscr{A}}_j = \begin{bmatrix} 0 & 1 & 0 \\ 0 & 0 & 1 \\ -k_{1,j} & -k_{2,j} & -k_{3,j} \end{bmatrix}, \quad 1 \le j \le 5,$$

Assume parameters and initial conditions that ensure chaotic behavior as in example 4 (Sect. 7.4.4). Let $k_{s,j} = \begin{bmatrix} 10 & 20 & 10 \end{bmatrix}$ for $1 \le j \le 5$ and diffusive coupling gains as $\kappa_1 = 2$, $\kappa_2 = 2$ and $\kappa_3 = 2$.

Figure 7.16 illustrates synchronization error convergence to the origin in transformed and original coordinates, respectively, FGMS is shown in Figs. 7.17 and 7.18 in transformed coordinates. And, taking the mapping (7.21) FGMS is shown in original coordinates in Figs. 7.19 and 7.20. The effectiveness of our approach can be verified from the multi-synchronization trajectories. Here, multi-synchronization is shown for two decoupled groups. Note that the synchronization error converges asymptotically to the origin. Convergence to the trajectories of master systems is unaffected from the interaction between the slave systems, this is due all eigenvalues of $(-\mathfrak{L} + \Xi)$ have negative real parts. It can be seen that any interaction is allowed, and in this example, first group have a undirected graph and second group have a directed graph. It is worth mentioning that the state of GS and CS can be achieved in the network, that is, the state of FGMS is present.

7.5 Concluding Remarks

In this chapter, we tackled the fractional generalized synchronization problem of non-linear commensurate fractional order Liouvillian systems, this problem was solved via multiple fractional dynamical feedbacks from a chain of fractional integrators inspired in differential algebra techniques. Moreover, we proposed a methodology for generalized synchronization of strictly different fractional order Liouvillian non-linear systems, we have designed a family of transformations by means of a family of fractional differential primitive elements and their fractional derivatives to carry out the families of master and slave systems to a MFGOCF, then generalized multi-synchronization is achieved. We have also considered the case of complex interaction between slave systems as a natural extension of Theorem 7.1, we showed that there is no restriction on the interplay between slave systems and synchronization error convergence to the origin. We introduced definitions related with the concept of fractional Liouvillian algebraic observability, Picard–Vessiot systems, and the concept of

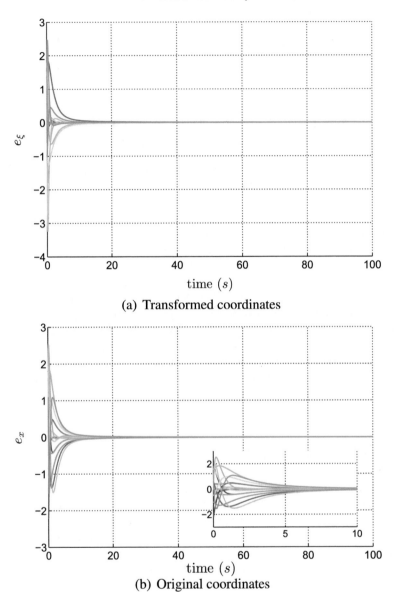

(a) Transformed coordinates

(b) Original coordinates

Fig. 7.16 Time evolution of synchronization error (see Sect. 7.4.4, Example 4)

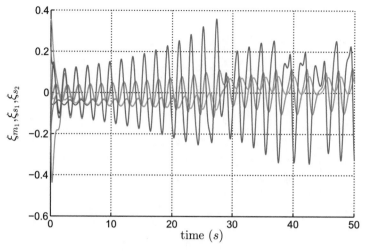

(a) Time evolution of synchronized trajectories

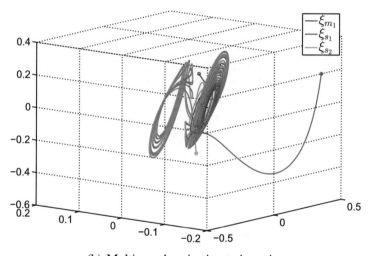

(b) Multi-synchronization trajectories

Fig. 7.17 FGMS in transformed coordinates (see Sect. 7.4.4, Example 4)

Fractional Generalized Multi-Synchronization (FGMS). And, we gave some numerical examples over a particular class of fractional order chaotic systems to show the effectiveness of this methodology. In our approach, we can explicitly obtain a fractional- order dynamical controller for the whole system instead of using a control signal for each fractional differential equation (active controllers), this versatility is obtained due to fractional differential primitive element. A disadvantage of the proposed approach is that we need a complete knowledge of the system to achieve

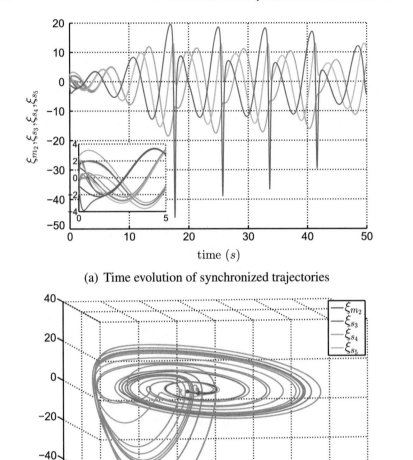

(a) Time evolution of synchronized trajectories

(b) Multi-synchronization trajectories

Fig. 7.18 FGMS in transformed coordinates (see Sect. 7.4.4, Example 4)

synchronization as is in active control case, but this problem is evidently solved with our methodology by means of observer-based fractional dynamical controllers. Potential applications of our current approach can be given in the context of secure communications and data encryption, our results can be naturally extended to this type of applications highlighting the Liouvillian feature that improve the reconstruction of the messages [18]. Forthcoming investigations will tackle the optimization of the proposed fractional controller as in [12], this will allow us to obtain an appropriate

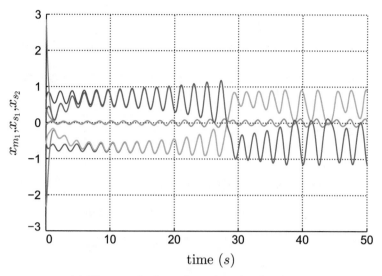

(a) Time evolution of synchronized trajectories

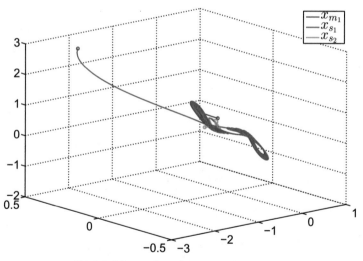

(b) Multi-synchronization trajectories

Fig. 7.19 FGMS in original coordinates (see Sect. 7.4.4, Example 4)

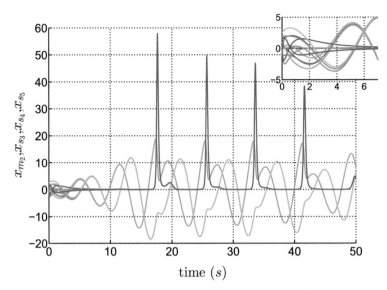

(a) Time evolution of synchronized trajectories

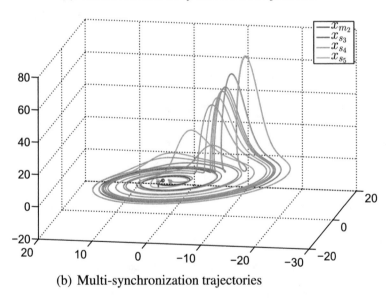

(b) Multi-synchronization trajectories

Fig. 7.20 FGMS in orignal coordinates (see Sect. 7.4.4, Example 4)

size of the control signal. Another direction is the extension to the case of parameter uncertainty, by designing fractional dynamical controllers in the sense of identifiable parameters [19, 20].

References

1. Florian Dörfler; Francesco Bullo. Synchronization in complex networks of phase oscillators: A survey. Automatica, 2014; 50(6): 1539–1564.
2. J. K. Hale, Diffusive coupling, dissipation, and synchronization, J. Dyn. Diff. Eqns., 1997; 9(1): 1–52.
3. Lin, D., & Wang, X., Observer-based decentralized fuzzy neural sliding mode control for interconnected unknown chaotic systems via network structure adaptation. Fuzzy Sets and Systems, 2010; 161(15), 2066–2080.
4. R. Martínez-Guerra, C. D. Cruz-Ancona, C. D., & C. A. Pérez-Pinacho, Generalized multi-synchronization viewed as a multi-agent leader-following consensus problem, Applied Mathematics and Computation, 2016; 282: 226–236.
5. Reza Olfati Saber, Alex Fax, Richard M. Murray, Consensus and Cooperation in Networked Multi-Agent Systems. Proceedings of the IEEE, 2007; 95(1): 215–233.
6. S. H. Strogatz, Exploring complex networks, Nature, 2001; 410(6825), 268–276.
7. Wang, Y., Zhang, H., Wang, X., & Yang, D., Networked synchronization control of coupled dynamic networks with time-varying delay. IEEE Transactions on Systems, Man, and Cybernetics, Part B (Cybernetics), 2010; 40(6), 1468–1479.
8. Wei Ren, Randal W. Beard and M. Atkins, Information Consensus in Multivehicle Cooperative Control, IEEE Control Systems Magazine, pp. 72–81 ,2007.
9. Delshad, S. S., Asheghan, M. M., & Beheshti, M. H., Synchronization of N-coupled incommensurate fractional-order chaotic systems with ring connection. Communications in Nonlinear Science and Numerical Simulation, 2011, 16(9): 3815–3824.
10. Wang, J., Ma, Q., Chen, A., & Liang, Z., Pinning synchronization of fractional-order complex networks with Lipschitz-type nonlinear dynamics. ISA transactions, 2015, 57: 111–116.
11. Wu, X., Lai, D., & Lu, H., Generalized synchronization of the fractional-order chaos in weighted complex dynamical networks with nonidentical nodes, Nonlinear Dynamics, 2012, 69(1): 667–683.
12. Behinfaraz, R., & Badamchizadeh, M. Optimal synchronization of two different incommensurate fractional-order chaotic systems with fractional cost function. Complexity, 2016, 21(S1): 401–416.
13. R. Martínez-Guerra, & J. L. Mata-Machuca, Fractional generalized synchronization in a class of nonlinear fractional order systems, Nonlinear Dynamics. 2014; 77(4): 1237–1244.
14. M.S. Tavazoei, M. Haeri, Synchronization of chaotic fractional-order systems via active sliding mode controller, Physica A: Statistical Mechanics and its Applications. 2008; 387(1): 57–70.
15. Wang, X. Y., Zhang, X., & Ma, C. Modified projective synchronization of fractional-order chaotic systems via active sliding mode control. Nonlinear Dynamics, 2012; 69(1): 511–517.
16. Wang, X. Y., He, Y. J., & Wang, M. J., Chaos control of a fractional order modified coupled dynamos system. Nonlinear Analysis: Theory, Methods & Applications, 2009, 71(12): 6126–6134.
17. I. Petrás, Fractional-Order Nonlinear Systems: Modeling, Analysis and Simulation, first ed., Springer, 2011.
18. R. Martínez-Guerra, J.J. Montesinos García & S.M. Delfin Prieto, Secure communications via synchronization of Liouvillian chaotic systems. Journal of the Franklin Institute, 2016; 353(17): 4384–4399.

19. C. Aguilar-Ibañez, R. Martínez-Guerra, R. Aguilar-López, J. L. Mata-Machuca, Synchronization and parameter estimations of an uncertain Rikitake system, Physics Letters A 08/2010; 374(35): 3625–3628.
20. R. Martínez-Guerra & C. D. Cruz-Ancona, Algorithms of Estimation for Nonlinear Systems: A Differential and Algebraic Viewpoint, Springer 2017.

Chapter 8
An Observer for a Class of Incommensurate Fractional Order Systems

8.1 Introduction

The main contribution in this chapter is to show a new observer for the synchronization problem in partially known nonlinear incommensurate fractional-order systems, we propose a novel technique using the master–slave synchronization scheme for estimating the unknown state variables based on a new IFAO property. As far as we know in the literature, this class of estimation scheme has not been used in incommensurate fractional-order systems.

The remainder of this chapter is organized as follows: In Sect. 8.2, we introduce the new concept given by Definition 2.10 (IFAO) as well as we propose a new system to estimate the unknown dynamics (slave system) so-called Incommensurate Fractional Reduced-Order Observer (IFROO), In Sect. 8.3 we apply our methodology to an Incommensurate Fractional-Order Rössler, Chua–Hartley and Financial Systems, the intention of choosing these systems is to clarify the proposed methodology and to highlight the simplicity and flexibility of the suggested approach, also we show some numerical results to confirm the effectiveness of the suggested approach, Finally, in Sect. 8.4, we close this chapter with some concluding remarks.

8.2 Problem Statement and Main Result

Now, consider the following class of incommensurate fractional order system:

$$\frac{d^{\alpha_i}}{dt^{\alpha_i}} x_i = f_i(x_1, \ldots, x_n), \, 1 \le i \le n, i \in \mathbb{Z}^+ \tag{8.1}$$

where α_i's are rational numbers between 0 and 1.

Consider the system given by (8.1), we will separate in two dynamical systems with states $\bar{x} \in \mathbb{R}^p$ and $\eta \in \mathbb{R}^{n-p}$, respectively, with $x_i^T = (\bar{x}_i^T, \eta_i^T)$ the first system

© Springer Nature Switzerland AG 2018
R. Martínez-Guerra and C. A. Pérez-Pinacho, *Advances in Synchronization of Coupled Fractional Order Systems*, Understanding Complex Systems,
https://doi.org/10.1007/978-3-319-93946-9_8

describes the known states and the second system represents unknown states, then the system (8.1) can be written as:

$$\bar{x}^{(\alpha_i)} = \bar{f}(\bar{x}, \eta)$$
$$\eta^{(\alpha_j)} = \Delta(\bar{x}, \eta) \qquad (8.2)$$
$$y_{\bar{x}} = h_{\bar{x}}(\bar{x})$$

where $\bar{x} \in \mathbb{R}^p$, $h : \mathbb{R}^p \to \mathbb{R}^q$ is a continuous function and $1 \leq p \leq n$ and $f^T(x) = (\bar{f}^T(\bar{x}, \eta), \Delta^T(\bar{x}, \eta))$, $\bar{f} \in \mathbb{R}^p$, $\Delta \in \mathbb{R}^{n-p}$.

Now, the problem is: How can we estimate the η's states? this question arises because if we know the η's states, we can use this signals to generate measuring depending on them.

If we assume that the components of unknown state vector η are IFAO (see Definition 2.10), then we can describe our problem in terms of the master–slave synchronization scheme, which is defined in the following way. Let us consider the master system:

$$\eta^{(\alpha_i)} = \Delta_i(\bar{x}, \eta) \qquad (8.3)$$

$$y_{\eta_i} = \eta_i = \phi_i(y_{\bar{x}}, y_{\bar{x}}^{(\alpha_1)}, D^{\alpha_2} D^{\alpha_1} y_{\bar{x}}, \ldots, D^{\alpha_n} \ldots D^{\alpha_1} y_{\bar{x}}) \qquad (8.4)$$

and consider an unknown dynamic:

$$\eta^{(\bar{\alpha}_i)} = \bar{\Delta}_i(\bar{x}, \eta) \qquad (8.5)$$

where $0 < \bar{\alpha}_i < 2$ is a rational number and $\bar{\Delta}_i(\bar{x}, \eta)$ is an unknown dynamics which contains $\Delta_i(\bar{x}, \eta)$. Now, let us propose an Incommensurate Fractional Reduced-Order Observer (IFROO) with order $\bar{\alpha}_i$, so the slave system is given by:

$$\hat{\eta}^{(\bar{\alpha}_i)} = k_{\hat{\eta}_i}(\eta_i - \hat{\eta}_i) \qquad (8.6)$$

$$y_{\hat{\eta}_i} = \hat{\eta}_i \qquad (8.7)$$

In the master–slave synchronization scheme, the output of the master system represents the target signal, while slave's output is the response signal. Therefore, given a master system and our slave system, it should be determined some conditions in order to synchronize the output of slave system with the output of master system.

Let us define the synchronization error as:

$$e_i = y_{\eta_i} - y_{\hat{\eta}_i} = \eta_i - \hat{\eta}_i \qquad (8.8)$$

We establish the following assumptions:

H1: η_i satisfies IFAO property for $i \in (p+1, \ldots, n)$
H2: $\bar{\Delta}_{p+1}$ is bounded, i.e., $\exists\ M \in \mathbb{R}^+$ such that $\|\bar{\Delta}(X)\| \leq M, \forall\ x$ in a compact set Ω, where $\bar{\Delta}(X) = (\bar{\Delta}_1, \ldots, \bar{\Delta}_n)^T$
H3: $k_{\hat{\eta}_i} \in \mathbb{R}^+$

Now, we are in position to propose the proposition

Proposition 8.1 *Let the system (8.1) which can be expressed as (8.2), where the above conditions are fulfilled, then (8.8) converges asymptotically to an compact set* $\bar{B}_r(0)$, *with* $r = \dfrac{M}{k_{\hat{\eta}_i}}$, *i.e., the synchronization is achieved (See Chap. 2).*

Proof From H1, we can write Eqs. (8.5)–(8.8). Taking the fractional derivative of the Eq. (8.8), we have:

$$e_i^{(\bar{\alpha}_i)} = \eta_i^{(\bar{\alpha}_i)} - \hat{\eta}_i^{(\bar{\alpha}_i)} \tag{8.9}$$

Substituting the fractional dynamics (8.5) and (8.6) into (8.8), we obtain:

$$e_i^{(\bar{\alpha}_i)} + k_{\hat{\eta}_i} e_i = \Delta_i\ (x) \tag{8.10}$$

There exists a unique solution for the system (8.10), due to $\Delta_i\ (x(t)) - k_{\hat{\eta}_i} e_i(t)$ is a Lipschitz continuous function on e.[1] Solving the above equation, we have:

$$e_i(t) = e_{i_0} E_{\bar{\alpha}_i,1}(-k_{\hat{\eta}_i} t^{(\bar{\alpha}_i)}) + \tag{8.11}$$
$$+ \int_0^t (t-\tau)^{\bar{\alpha}_i - 1} E_{\bar{\alpha}_i,\bar{\alpha}_i}(k_{\hat{\eta}_i}(t-\tau)^{\bar{\alpha}_i})\ \Delta_i\ (\tau) d\tau$$

where $e_i(0) = e_{i_0}$. Using Triangle and Cauchy–Schwarz inequalities and H2

$$|\ e_i(t)\ | \leqslant |\ e_{i_0} E_{\bar{\alpha}_i,1}(-k_{\hat{\eta}_i} t^{\bar{\alpha}_i})\ | \tag{8.12}$$
$$+ M \int_0^t |\ (t-\tau)^{\bar{\alpha}_i - 1} E_{\bar{\alpha}_i,\bar{\alpha}_i}(-k_{\hat{\eta}_i}(t-\tau)^{\bar{\alpha}_i})\ |\ d\tau$$

The functions $(t-\tau)^{\bar{\alpha}_i - 1} E_{\bar{\alpha}_i,\bar{\alpha}_i}(-k_{\hat{\eta}_i}(t-\tau)^{\bar{\alpha}_i})$ and $E_{\bar{\alpha}_i}(-k_{\hat{\eta}_i} t^{\bar{\alpha}_i})$ are not negative due to Property 2.1 in Chap. 2 of Mittag-Leffler function and H3.

$$|\ e_i(t)\ | \leqslant |\ e_{i_0}\ |\ E_{\bar{\alpha}_i,1}(-k_{\hat{\eta}_i} t^{\bar{\alpha}_i}) \tag{8.13}$$
$$+ M \int_0^t (t-\tau)^{\bar{\alpha}_i - 1} E_{\bar{\alpha}_i,\bar{\alpha}_i}(-k_{\hat{\eta}_i}(t-\tau)^{\bar{\alpha}_i}) d\tau$$

---

[1]Equation (8.10) is nonautonomous, but the Lipschitz condition assures a unique solution [1].

Using Property 2.2 in Chap. 2 of Mittag-Leffler function

$$| e_i(t) | \leqslant | e_{i_0} | E_{\bar{\alpha}_i, 1}(-k_{\hat{\eta}_i} t^{\bar{\alpha}_i}) + M t^{(\bar{\alpha}_i)} E_{\bar{\alpha}_i, \bar{\alpha}_i+1}(-k_{\hat{\eta}_i} t^{\bar{\alpha}_i}) \tag{8.14}$$

If $t \rightarrow \infty$, we use the Eq. (8.4) with $\mu = 3\pi \dfrac{\bar{\alpha}_i}{4}$ due to H3.

$$\lim_{t \to \infty} | e_i(t) | \leq | e_{i_0} | \lim_{t \to \infty} E_{\bar{\alpha}_i, 1}(-k_{\hat{\eta}_i} t^{\bar{\alpha}_i}) \tag{8.15}$$

$$+ M \lim_{t \to \infty} t^{\bar{\alpha}_i} E_{\bar{\alpha}_i, \bar{\alpha}_i+1}(-k_{\hat{\eta}_i} t^{\bar{\alpha}_i}) = \frac{M}{k_{\hat{\eta}_i}} \qquad \square$$

8.3   Numerical Results

In this section, we study the problem of synchronization for incommensurate fractional dynamical systems by means of numerical simulations. Consider the fractional-order Rössler system [2] as follows:

$$x_1^{(\alpha_1)} = - x_1 - x_3$$
$$x_2^{(\alpha_2)} = x_1 + 0.63x_2 \tag{8.16}$$
$$x_3^{(\alpha_3)} = 0.2 + x_3(x_1 - 10)$$

where $x = (x_1, x_2, x_3)^T$ is the state vector, $\alpha_1 = 0.9$, $\alpha_2 = 0.8$ and $\alpha_3 = 0.7$ and we take the system output as $y = x_2$. The system (8.16) can be rewritten as (8.33):

$$\bar{x}_2^{(\alpha_2)} = \eta_1 + 0.63\bar{x}_2$$
$$\eta_1^{(\alpha_1)} = - \bar{x}_2 - \eta_3 \tag{8.17}$$
$$\eta_3^{(\alpha_3)} = 0.2 + \eta_3(\eta_1 - 10)$$

where $x_2 = \bar{x}_2, x_1 = \eta_1, x_3 = \eta_3$ and $y = \bar{x}_2$. From (8.17) the following relationships are achieved:

$$\eta_1 = y^{(\alpha_2)} - 0.63y \tag{8.18}$$

$$\eta_3 = -y + 0.63y^{(\alpha_1)} - D^{\alpha_2} D^{\alpha_1} y \tag{8.19}$$

from (8.18) and (8.19) we can see that $\eta_3 = x_3$ and $\eta_1 = x_1$ are IFAO;
    The master systems are given by:

$$\eta_1^{(\alpha_1)} = -\bar{x}_2 - \eta_3 \tag{8.20}$$

$$y_{\eta_1} = \eta_1 = y^{(\alpha_2)} - 0.63y \tag{8.21}$$

$$\eta_3^{(\alpha_3)} = 0.2 + \eta_3(\eta_1 - 10) \tag{8.22}$$

$$y_{\eta_3} = \eta_3 = -y + 0.63y^{(\alpha_1)} - D^{\alpha_2}D^{\alpha_1}y \tag{8.23}$$

Remark 8.1 To design the slave system, the derivative order for each $\hat{\eta}_i$ is chosen such that $\bar{\alpha}_i$ is equal to the biggest fractional-order derivative of the output y.

Now, we design the slave system for (8.21) in this case $\bar{\alpha}_i = \alpha_2 = 0.8$, then we have:

$$\hat{\eta}_1^{(\alpha_2)} = k_1(\eta_1 - \hat{\eta}_1) \tag{8.24}$$

and substituting (8.18) into (8.24) we obtain:

$$\hat{\eta}_1^{(\alpha_2)} = k_1(y^{(\alpha_2)} - 0.63y - \hat{\eta}_1) \tag{8.25}$$

In order to avoid fractional-order derivatives, it is proposed a change of variable $\hat{\eta}_1 = \gamma_1 + k_1 y$ and from Eq. (8.25) after some manipulations, we obtain:

$$\gamma_1^{(\alpha_2)} = k_1(-0.63y - \gamma_1 - k_1 y) \tag{8.26}$$

Now, it is time to estimate x_3, but in this case we consider x_1 unknown, then we use:

$$x_1^{(\alpha_1)} = D^{\alpha_2}D^{\alpha_1}x_2 - 0.63x_2^{(\alpha_1)} \tag{8.27}$$

Due to the system is incommensurate, previous equation shows a derivative which depends on different fractional orders; however, it is possible to obtain a reduced order observer after additional manipulations as it is shown. From the first equation of (8.16) and Eq. (8.27), the following expression is obtained:

$$x_3 = \eta_3 = -y + 0.63y^{(\alpha_1)} - D^{\alpha_2}D^{\alpha_1}y \tag{8.28}$$

In this case $\bar{\alpha}_i = \alpha_1 + \alpha_2 = 1.7$ which is smaller than 2. At the beginning, the slave system for (8.22) has the following representation:

$$D^{\alpha_2}D^{\alpha_1}\hat{\eta}_3 = k_2(-y - D^{\alpha_2}D^{\alpha_1}y + 0.63y^{(\alpha_1)} - \hat{\eta}_3) \tag{8.29}$$

We introduce in (8.29) the change of variable $\hat{\eta}_3 = \beta_1 - k_2 y$ in order to avoid the term $y^{(\alpha_1+\alpha_2)}$, then we have:

$$D^{\alpha_2}D^{\alpha_1}\beta_1 = k_2(-y + 0.63y^{(\alpha_1)} - \beta_1 + k_2 y) \tag{8.30}$$

Finally, we need to avoid one more term, that is $y^{(\alpha_1)}$. To achieve this goal, we propose a change of variable as follows: first consider a new variable β_2 and substituting the change of variable $\beta_1 = \beta_2^{(-\alpha_2)} + 0.63k_2 y^{(-\alpha_2)}$ into (8.30); then, after some algebraic manipulations it is possible to achieve the following relationship:

$$\beta_2^{(\alpha_1)} = k_2(-y - \beta_2^{(-\alpha_2)} - 0.63 k_2 y^{(-\alpha_2)} + k_2 y) \tag{8.31}$$

We select the observer's constant parameters as $k_1 = 120$ and $k_2 = 7000$.

Figures 8.1 and 8.2 show the original system states and the slave system synchronized with the master system respectively. To end this example, the Figs. 8.3 and 8.4 evince the convergence of estimates to original states. Now, consider the fractional-order Chua system [3] as follows:

$$
\begin{aligned}
x_1^{(\alpha_1)} &= ax_2 + \frac{ax_1}{7} - \frac{2ax_1^3}{7} \\
x_2^{(\alpha_2)} &= x_1 - x_2 + x_3 \\
x_3^{(\alpha_3)} &= -\beta x_2
\end{aligned}
\tag{8.32}
$$

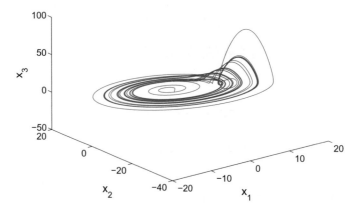

Fig. 8.1 Phase plot of the incommensurate fractional-order system with initial conditions $x_1(0) = 1$, $x_2 = 0$ and $x_3(0) = -5$

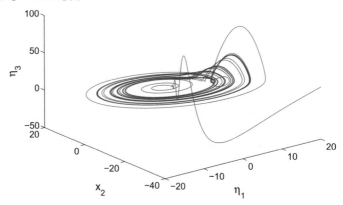

Fig. 8.2 Phase plot of the slave incommensurate fractional-order system with initial conditions $\hat{\eta}_1(0) = 100$ and $\hat{\eta}_3(0) = 200$

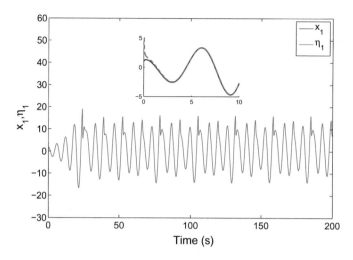

Fig. 8.3 Synchronization of the incommensurate fractional-order system, state x_1 versus estimate η_1

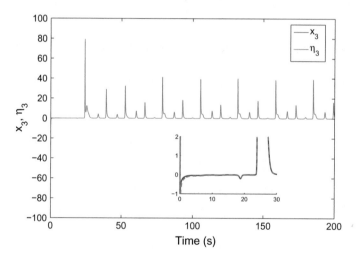

Fig. 8.4 Synchronization of the incommensurate fractional-order system, state x_3 versus estimate η_3

where $a = 12.75$, $\beta = \dfrac{100}{7}$, $x = (x_1, x_2, x_3)^T$ is the state vector, $\alpha_1 = 0.99$, $\alpha_2 = 0.91$ and $\alpha_3 = 0.95$ and we take the system output as $y = x_1$. The system (8.32) can be rewritten as:

$$
\begin{aligned}
\bar{x}_1^{(\alpha_1)} &= a\eta_2 + \frac{a\bar{x}_1}{7} - \frac{2a\bar{x}_1^3}{7} \\
\eta_2^{(\alpha_2)} &= \bar{x}_1 - \eta_2 + \eta_3 \\
\eta_3^{(\alpha_3)} &= -\beta\eta_2
\end{aligned}
\tag{8.33}
$$

where $x_1 = \bar{x}_2$, $x_2 = \eta_2$, $x_3 = \eta_3$ and $y = \bar{x}_1$. From (8.33), the following relationships are obtained:

$$\eta_2 = \frac{y^{(\alpha_1)}}{a} - \frac{y}{7} + \frac{2y^3}{7} \tag{8.34}$$

$$\eta_3 = \frac{D^{\alpha_2} D^{\alpha_1} y}{a} - \frac{y^{(\alpha_2)}}{7} + \frac{2y^{3(\alpha_2)}}{7} - y + \frac{y^{(\alpha_1)}}{a} - \frac{y}{7} + \frac{2y^3}{7} \tag{8.35}$$

from (8.34) and (8.35), we can see that $\eta_2 = x_2$ and $\eta_3 = x_3$ are IFAO;

$$\eta_2^{(\alpha_2)} = \bar{x}_1 - \eta_2 + \eta_3 \tag{8.36}$$

$$y_{\eta_2} = \eta_2 = \frac{y^{(\alpha_1)}}{a} - \frac{y}{7} + \frac{2y^3}{7} \tag{8.37}$$

$$\eta_3^{(\alpha_3)} = -\beta\eta_2 \tag{8.38}$$

$$y_{\eta_3} = \eta_3 = \frac{D^{\alpha_2} D^{\alpha_1} y}{a} - \frac{y^{(\alpha_2)}}{7} + \frac{2y^{3(\alpha_2)}}{7} - y + \frac{y^{(\alpha_1)}}{a} - \frac{y}{7} + \frac{2y^3}{7} \tag{8.39}$$

Now, we design the slave system for (8.21) in this case $\bar{\alpha}_i = \alpha_1 = 0.99$, then we have:

$$\hat{\eta}_2^{(\alpha_1)} = k_1(\eta_2 - \hat{\eta}_2) \tag{8.40}$$

and substituting (8.34) into (8.40) we obtain:

$$\hat{\eta}_2^{(\alpha_1)} = k_1\left(\frac{y^{(\alpha_1)}}{a} - \frac{y}{7} + \frac{2y^3}{7} - \hat{\eta}_2\right) \tag{8.41}$$

In order to avoid fractional-order derivatives, it is proposed a change of variable $\hat{\eta}_2 = \gamma_1 + \frac{k_1 y}{a}$ and from Eq. (8.41) after some algebraic manipulations we obtain:

$$\gamma_1^{(\alpha_1)} = k_1\left(-\frac{y}{7} + \frac{2y^3}{7} - \gamma_1 - \frac{k_1 y}{a}\right) \tag{8.42}$$

Now, we estimate x_3, but in this case we consider x_2 unknown, then we use:

$$x_2^{(\alpha_2)} = \bar{x}_1 - \eta_2 + \eta_3 \tag{8.43}$$

Since the system is incommensurate, previous equation shows a derivative which depends on different fractional orders; however, it is possible to obtain a reduced order observer after some manipulations as it is shown. From the first equation of (8.32) and Eq. (8.43), the following expression is obtained:

$$x_3 = \eta_3 = \eta_2^{(\alpha_2)} - y - \eta_2$$

$$\eta_3 = \frac{D^{\alpha_2} D^{\alpha_1} y}{a} - \frac{y^{(\alpha_2)}}{7} + \frac{2y^{3(\alpha_2)}}{7} - y + \frac{y^{(\alpha_1)}}{a} - \frac{y}{7} + \frac{2y^3}{7} \tag{8.44}$$

In this case $\bar{\alpha}_i = \alpha_1 + \alpha_2 = 1.9$ which is smaller than 2. At the beginning, the slave system for (8.38) has the following representation:

$$\hat{\eta}_3^{(\alpha_1 + \alpha_2)} = k_2(\eta_3 - \hat{\eta}_3) \tag{8.45}$$

$$\hat{\eta}_3^{(\alpha_1 + \alpha_2)} = k_2 \left(\frac{y^{(\alpha_1 + \alpha_2)}}{a} - \frac{y^{(\alpha_2)}}{7} + \frac{2y^{3(\alpha_2)}}{7} \right.$$
$$\left. - y + \frac{y^{(\alpha_1)}}{a} - \frac{y}{7} + \frac{2y^3}{7} - \hat{\eta}_3 \right) \tag{8.46}$$

We introduce in (8.46) the change of variable $\hat{\eta}_3 = \beta_1 + \dfrac{k_2 y}{a}$ in order to avoid the term $D^{\alpha_2} D^{\alpha_1} y$, then we have:

$$\beta_1^{(\alpha_1 + \alpha_2)} = k_2 \left(-\frac{y^{(\alpha_2)}}{7} + \frac{2y^{3(\alpha_2)}}{7} - y + \frac{y^{(\alpha_1)}}{a} \right.$$
$$\left. - \frac{y}{7} + \frac{2y^3}{7} - \beta_1 - \frac{k_2 y}{a} \right) \tag{8.47}$$

Finally, we need to avoid derivatives $y^{(\alpha_1)}$ and $y^{(\alpha_2)}$. To achieve this goal, we propose a change of variable as follows:

$$\beta_1 = \beta_2^{(-\alpha_2)} - \frac{k_2 y^{(-\alpha_1)}}{7} + \frac{2k_2 y^{3(-\alpha_1)}}{7} + \frac{k_2 y^{(-\alpha_2)}}{a} \tag{8.48}$$

Substituting the change (8.48) into (8.30) then, after some algebraic manipulations, finally, we have the next:

$$\beta_2^{(\alpha_1)} = k_2 \left(-y \frac{-y}{7} + \frac{2y^3}{7} - \frac{k_2 y}{a} - \beta_1 \right) \tag{8.49}$$

We select the observer's constant parameters as $k_1 = 100$ and $k_2 = 1000$.

In Fig. 8.5, we can observe the original systems while Fig. 8.6 shows the slave system synchronized with the master system. Finally, the convergence of estimates to original states it is shown in Figs. 8.7 and 8.8.

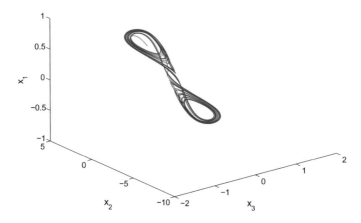

Fig. 8.5 The incommensurate fractional-order system with initial conditions $x_1(0) = 1$, $x_2 = 2$ and $x_3(0) = -1$

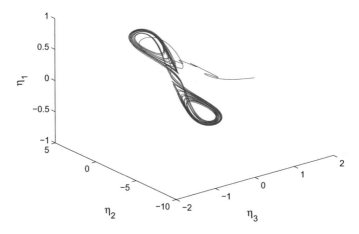

Fig. 8.6 The slave incommensurate fractional-order system with initial conditions $\hat{\eta}_1(0) = -10$ and $\hat{\eta}_2(0) = 0$

Finally, we study the problem of fractional financial system [4]

$$
\begin{aligned}
x_1^{(\alpha_1)} &= x_3 + (x_2 - 3)x_1 \\
x_2^{(\alpha_2)} &= 1 - 0.1x_2 - x_1^2 \\
x_3^{(\alpha_3)} &= -x_1 - x_3
\end{aligned}
\tag{8.50}
$$

where the interest rate, investment demand, and price index are given by x_1, x_2, x_3, respectively, $x = (x_1, x_2, x_3)^T$ is the state vector, $\alpha_1 = 0.95$, $\alpha_2 = 0.98$, and $\alpha_3 = 0.99$, and we take the system output as $y = x_3$. The system (8.50) can be rewritten as (8.33):

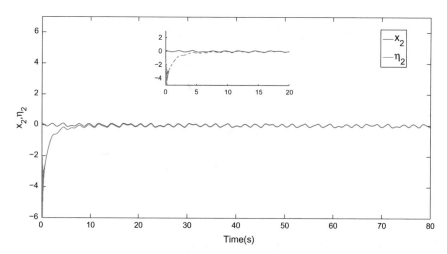

Fig. 8.7 Synchronization of the incommensurate fractional-order system, state x_2 versus estimate η_2

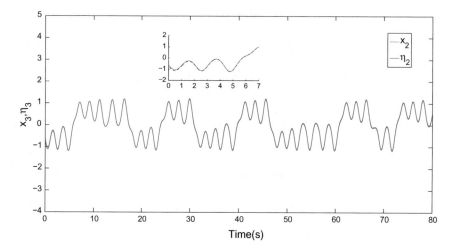

Fig. 8.8 Synchronization of the incommensurate fractional-order system, state x_3 versus estimate η_3

$$\bar{x}_3^{(\alpha_3)} = -\eta_1 - \bar{x}_3$$
$$\eta_1^{(\alpha_1)} = \bar{x}_3 + (\eta_2 - 3)\eta_1 \qquad (8.51)$$
$$\eta_2^{(\alpha_2)} = 1 - 0.1\eta_2 - \eta_1^2$$

where $x_3 = \bar{x}_3$, $x_1 = \eta_1$, $x_2 = \eta_2$ and $y = \bar{x}_3$. From (8.51) the following relations are achieved:

$$\eta_1 = -y^{(\alpha_3)} - y \qquad (8.52)$$

$$\eta_2 = \frac{1}{-y^{(\alpha_3)} - y}[y^{(\alpha_1 + \alpha_3)} - y^{(\alpha_1)} - y + 3(-y^{(\alpha_3)} - y)] \tag{8.53}$$

from (8.52) and (8.53), we can see that $\eta_1 = x_1$ and $\eta_2 = x_2$ are IFAO.

$$\eta_1^{(\alpha_1)} = \bar{x}_3 + (\eta_2 - 3)\eta_1 \tag{8.54}$$

$$y_{\eta_1} = \eta_1 = -y^{(\alpha_3)} - y \tag{8.55}$$

$$\eta_2^{(\alpha_2)} = 1 - 0.1\eta_2 - \eta_1^2 \tag{8.56}$$

$$y_{\eta_2} = \frac{1}{-y^{(\alpha_3)} - y}[y^{(\alpha_1 + \alpha_3)} - y^{(\alpha_1)} - y + 3(-y^{(\alpha_3)} - y)] \tag{8.57}$$

Now, we design the slave system for (8.54), then we have:

$$\hat{\eta}_1^{(\alpha_3)} = k_1(\eta_1 - \hat{\eta}_1) \tag{8.58}$$

and substituting (8.52) into (8.58) we obtain:

$$\hat{\eta}_1^{(\alpha_3)} = k_1(-y^{(\alpha_3)} - y - \hat{\eta}_1) \tag{8.59}$$

In order to avoid fractional-order derivatives, it is proposed a change of variable $\hat{\eta}_1 = \gamma_1 - k_1 y$ and from Eq. (8.59) after some manipulations we obtain:

$$\gamma^{(\alpha_3)} = k_1(k_1 y - y - \gamma_1) \tag{8.60}$$

To estimate x_2 there is a problem when $-y^{(\alpha_3)} - y = 0$, in this moment the IFAO property is lost so, in order to overcome this drawback, from (8.51), we use as an estimate:

$$\eta_2 = 10 - 10\hat{\eta}_2^{(\alpha_2)} - 10\hat{\eta}_1^2 \tag{8.61}$$

then, the slave system for (8.56) is given by:

$$\hat{\eta}_2^{(\alpha_2)} = k_2(\eta_2 - \hat{\eta}_2) \tag{8.62}$$

substituting (8.61) into (8.62) and after some algebraic manipulations, we achieve the following observer:

$$\hat{\eta}_2^{(\alpha_2)} = \frac{1}{1 + 10k_2}(10k_2 - 10k_2\hat{\eta}_1^2 - k_2\hat{\eta}_2) \tag{8.63}$$

Finally, the simulations shows the effectiveness of the proposed observer, in simulations the gains are $k_1 = 100$ and $k_2 = 100$.

Fig. 8.9 Original system
initial conditions
$x_0 = (2, 3, 2)$

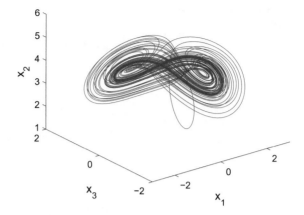

Fig. 8.10 Slave system
initial conditions $\eta_1 = 20$,
$\eta_2 = 30$

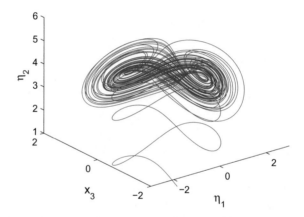

Fig. 8.11 State x_1 versus
estimate η_1

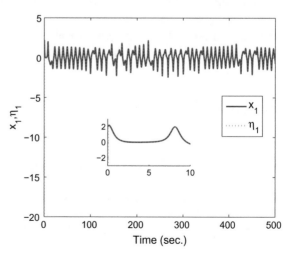

Fig. 8.12 State x_2 versus estimate η_2

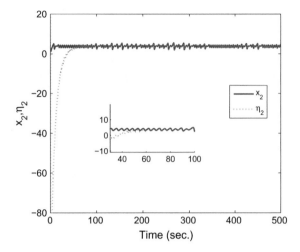

The Fig. 8.9 shows the original systems while Fig. 8.10 shows the slave system synchronized with the master system. To end this section, the Figs. 8.11 and 8.12 evince the convergence of estimates to original states.

8.4   Conclusions

In this chapter, it was introduced a new concept of Incommensurate Fractional Algebraic Observability (IFAO), we introduced a new observer (IFROO) to solve the synchronization problem for incommensurate fractional dynamical systems. The scheme was applied to incommensurate fractional chaotic systems; however, it could be applied to other classes of systems which satisfy the Proposition 8.1. Finally, numerical simulations showed the effectiveness of the suggested approach.

References

1. A.A. Kilbas, H.M. Srivastava, J.J. Trujillo, Theory and Applications of Fractional Differential Equations, Elsevier B.V., (2006).
2. Mohammad Saleh Tavazoei, Mohammad Haeri, Chaotic attractors in incommensurate fractional order systems, Physica D 237, 2628–2637, (2008).
3. R. Caponetto,G. Dongola, L. Fortuna,I. Petrás, Fractional Order Systems,World Scientific Series on Nonlinear Science, Series A. Vol. 72, 53–75, (2010).
4. W.C. Chen, Nonlinear dynamics and chaos in a fractional-order financial system, Chaos, Solitons and Fractals, 36 (5), 1305–1314, (2008).

Chapter 9
Fractional Generalized Quasi-synchronization of Incommensurate Fractional Order Oscillators

9.1 Introduction

GS was introduced in [1], but here definitions are extended and given in our own conception, for fractional order nonlinear systems, by using the fractional incommensurate differential primitive element. The problem of the Fractional Generalized Synchronization (FGS) was studied for a class of strictly different nonlinear commensurate fractional order systems in the master–slave configuration scheme [2]. Recently, numerous works have been reported on the problem of synchronization for incommensurate fractional order chaotic systems [3–6]. In general, study synchronization of strictly different systems is equivalently to study the asymptotic stability of the origin of the synchronization error or the stability of the synchronization manifold if possible. In many of these references, the stability of the incommensurate fractional order dynamics of the synchronization error is translated into a problem of stability of a commensurate fractional order or even an integer order system through a change of variable. In this chapter, we will show a convergence analysis directly from the incommensurate fractional order dynamics of the synchronization error. It is natural to present the incommensurate fractional order dynamics of the synchronization error in a modal decomposition due to each dynamics have different fractional order. Thus, we can obtain asymptotic convergence in a compact region near the origin in case of synchronization error for generalized synchronization of strictly different incommensurate fractional- order systems by using dynamical controllers obtained from differential algebraic techniques. In this chapter, the main contribution is a Fractional Generalized Synchronization constructive method for nonlinear incommensurate fractional order chaotic systems in a master–slave topology, this phenomena is studied from an algebraic and differential point of view, that allows us to construct an Incommensurate Fractional Generalized Observability Canonical Form (IFGOCF) from an adequate selection of a fractional differential primitive element and moreover, give explicit the form of the synchronization algebraic manifold for strictly different fractional order nonlinear systems. The former enables us

© Springer Nature Switzerland AG 2018
R. Martínez-Guerra and C. A. Pérez-Pinacho, *Advances in Synchronization of Coupled Fractional Order Systems*, Understanding Complex Systems, https://doi.org/10.1007/978-3-319-93946-9_9

to design an incommensurate fractional order dynamical controller able to achieve synchronization of strictly different incommensurate fractional order chaotic systems. Moreover, we introduce the concepts so-called Incommensurate Fractional Algebraic Observability and a fractional order Picard–Vessiot system. As far as we know, synchronization of strictly different incommensurate fractional order systems have not been tackled from this perspective. The rest of the chapter is organized as follows: A solution for the problem of generalized synchronization for incommensurate fractional order chaotic systems is shown in Sect. 9.2. In Sect. 9.3, the proposed methodology for fractional generalized synchronization between Chua–Hartley and Rössler is applied and numerical results are showed to confirm the effectiveness of the suggested approach. Finally, in Sect. 9.4, some concluding remarks are given.

9.2   Problem Statement and Main Result

Consider that there exists an element $y \in \mathbb{R}$ such that $(D^{\alpha_n} \cdots D^{\alpha_0} y)$ is analytically dependent on $(y, D^{\alpha_1} y, D^{\alpha_2} D^{\alpha_1} y, \ldots, D^{\alpha_{n-1}} \cdots D^{\alpha_1} y)$:

$$\bar{H}(y, D^{\alpha_1} y, D^{\alpha_1} D^{\alpha_2} y, \ldots, D^{\alpha_n} \cdots D^{\alpha_1} y, u, D^{\alpha_1} u, D^{\alpha_2} D^{\alpha_1} u, \ldots, D^{\alpha_n} \cdots D^{\alpha_1} u) \quad (9.1)$$

The system (9.1) can be solved locally as:

$$D^{\alpha_n} \cdots D^{\alpha_1} y = (y, D^{\alpha_1} y, D^{\alpha_1} D^{\alpha_2} y, \ldots, D^{\alpha_{n-1}} \cdots D^{\alpha_1}, u, D^{\alpha_1} u,$$
$$D^{\alpha_2} D^{\alpha_1} u, \ldots, D^{\alpha_{n-1}} \cdots D^{\alpha_1} u) + (D^{\alpha_n} \cdots D^{\alpha_1} u)$$

Then, we can establish the following lemma.

Lemma 9.1 *A nonlinear incommensurate fractional order system (2.21) is transformable to a IFGOCF if and only if it is PV.*

Proof (Sufficiency) Let the set $\{\varepsilon, D^{\alpha_1}, D^{\alpha_2} D^{\alpha_1} \varepsilon, \ldots, D^{\alpha_{n-1}} \cdots D^{\alpha_1} \varepsilon\}$ be a finite basis (PV) where $n \geq 0$ is the minimum integer such that $(D^{\alpha_n} \cdots D^{\alpha_1} y)$ is dependent on $y, D^{\alpha_1} y, D^{\alpha_2} D^{\alpha_1} y, \ldots, D^{\alpha_{n-1}} \cdots D^{\alpha_1}, u, D^{\alpha_1} u, D^{\alpha_2} D^{\alpha_1} u, \ldots, D^{\alpha_n} \cdots D^{\alpha_1} u$ Redefining $\xi_i = D^{\alpha_{i-1}} \cdots D^{\alpha_0} \varepsilon$, $1 \leq i \leq n$, this yields to next IFGOCF

$$\xi_j^{(\alpha_j)} = \xi_{j+1}, \quad 1 \leq j \leq n-1$$
$$\xi_n^{(\alpha_n)} = -\mathscr{L}(\xi_1, \ldots, \xi_n, u, u^{(\alpha_1)}, \ldots, D^{\alpha_{n-1}} \cdots D^{\alpha_1} u) + D^{\alpha_n} \cdots D^{\alpha_1} u$$
$$\bar{y} = \xi_1$$

(Necessity) It is immediate.   □

Now, consider a master–slave configuration of two incommensurate fractional order systems, assume the master system has the following form:

$$x_M^{(\alpha)} = F_M(x_M, u_M)$$
$$y_M = h_M(x_M, u_M) \tag{9.2}$$

and the slave system is given by:

$$x_S^{(\bar{\alpha})} = F_S(x_S, u_S(x_S, y_M))$$
$$y_S = h_S(x_S, u_S(x_S, y_M)) \tag{9.3}$$

where $x_S \in \mathbb{R}^n$, $F_S : \mathbb{R}^n \times \mathbb{R}^{\bar{m}_S} \to \mathbb{R}^n$, $F_M : \mathbb{R}^n \times \mathbb{R}^{\bar{m}_M} \to \mathbb{R}^n$, $x_M \in \mathbb{R}^n$, $h_S : \mathbb{R}^n \times \mathbb{R}^{\bar{m}_S} \to \mathbb{R}$, $h_M : \mathbb{R}^n \times \mathbb{R}^{\bar{m}_M} \to \mathbb{R}$, $u_M \in \mathbb{R}^{\bar{m}_M}$, $u_S \in \mathbb{R}^{\bar{m}_S}$, $u_S : \mathbb{R}^n \times \mathbb{R} \to \mathbb{R}^{\bar{m}_S}$, y_M, $y_S \in \mathbb{R}$, F_S, F_M, h_S, h_M are assumed to be polynomial in their arguments, with initial conditions $x_{M0} = x_M(0)$ and $x_{S0} = x_S(0)$ and assume the fractional order $\bar{\alpha}$ ($\bar{\alpha} = [\bar{\alpha}_1, \bar{\alpha}_2, \ldots, \bar{\alpha}_n]^T$ for $0 = \bar{\alpha}_0 < \bar{\alpha}_i < 2$, $i = 1, 2, \ldots, n$) is not necessarily equal to α.

In this case of strictly different Incommensurate Fractional Order Chaotic Systems, the well-understood definition of Fractional Generalized Synchronization lacks sense.

From Lemma 9.1, it is clear that we can obtain a corresponding IFGOCF for master and slave systems, and from these canonical forms, it is possible to obtain an incommensurate dynamical controller to ensure the state of FGS, this is summarized in the following Theorem and the main result of this chapter.

Theorem 9.1 *Let systems (9.2) and (9.3) be transformable to a IFGOCF. Let us define* $z_M = (z_{M_1}, z_{M_2}, \ldots, z_{M_n})^T$ *and* $z_S = (z_{S_1}, z_{S_2}, \ldots, z_{S_n})^T$ *as the trajectories of master and slave systems in the transformed coordinates, respectively, with* $z_{M_i} = D^{\alpha_{i-1}} \cdots D^{\alpha_1} y_M$ *and* $z_{S_i} = D^{\bar{\alpha}_{i-1}} \cdots D^{\bar{\alpha}_1} y_S$ *for* $1 \leq i \leq n$. *Moreover, assuming that H1 and H2 are fulfilled. Then, the solution of (9.4) converges asymptotically to a compact set* $\bar{B}_r(0)$, *with* $r = \max_{1 \leq i \leq n} \bar{r}_i$ *and* $\bar{r}_i = \sum_{l=1}^{n} |\lambda_l|^{i-1} \dfrac{\bar{g}_l}{|\lambda_l|}$. *In other words, generalized synchronization is achieved.*

Proof Without loss of generality suppose $u_m = 0 \in \mathbb{R}^{\bar{m}_n}$, and take the fractional incommensurate differential primitive element for the master system as:

$$y_M = \sum_i \alpha_{M_i} x_{M_i} =: z_{M_i}$$

where α_{M_i} is a differential quantities of u_M and their time finite fractional derivatives and for the slave system is:

$$y_S = \sum_i \alpha_{S_i} x_{S_i} + \sum_i \beta_{S_i} u_{S_i} =: z_{S_i},$$

where α_{S_i}, β_{S_i} are differential quantities of u_S and their time-fractional derivatives.

According to Lemma 9.1, we obtain the IFGOCF of system (9.2)

$$z_{M_j}^{(\alpha_j)} = z_{M_{j+1}}, \quad 1 \le j \le n-1$$
$$z_{M_n}^{(\alpha_n)} = -\mathscr{L}_M(z_{M_1}, z_{M_2}, \dots, z_{M_n})$$

and the IFGOCF of system (9.3),

$$z_{S_j}^{(\bar\alpha_j)} = z_{S_{j+1}}, \quad 1 \le j \le n-1$$
$$z_{S_n}^{(\bar\alpha_n)} = -\mathscr{L}_s(z_{S_1}, z_{S_2}, \dots, z_{S_n}, u_1, u_2, \dots, u_\gamma) + u_\gamma^{(\bar\alpha_\gamma)}$$

where

$$u_1 = u_S$$
$$u_2 = u_S^{(\bar\alpha_1)}$$
$$u_3 = D^{\bar\alpha_2} D^{\bar\alpha_1} u_S$$
$$\vdots$$
$$u_\gamma = D^{\bar\alpha_{\gamma-1}} \cdots D^{\bar\alpha_1} u_S$$

Consider the dynamics of the synchronization error $e_z = z_M - z_S$ given by the system:

$$e_{z_j}^{(\bar\alpha_j)} = e_{z_{j+1}} + \mathscr{G}_j(z_{M_{j+1}}, z_{M_j}^{(\bar\alpha_j)}) \quad 1 \le j \le n-1$$
$$e_{z_n}^{(\bar\alpha_n)} = z_{M_n}^{(\bar\alpha_n)} + \mathscr{L}_s(z_{S_1}, z_{S_2}, \dots, z_{S_n}, u_1, u_2, \dots, u_\gamma) - u_\gamma^{(\bar\alpha_n)}$$

where $\mathscr{G}_j(z_{M_{j+1}}, z_{M_j}^{(\bar\alpha_j)}) := z_{M_j}^{(\bar\alpha_j)} - z_{M_{j+1}}$ for $1 \le j \le n-1$.

Then, we can rewrite the error dynamics as an augmented system:

$$e_{z_j}^{(\bar\alpha_j)} = e_{z_{j+1}} + \mathscr{G}_j(z_{M_{j+1}}, z_{M_j}^{(\bar\alpha_j)}) \quad 1 \le j \le n-1$$
$$e_{z_n}^{(\bar\alpha_n)} = z_{M_n}^{(\bar\alpha_n)} + \mathscr{L}_s(z_{S_1}, z_{S_2}, \dots, z_{S_n}, u_1, u_2, \dots, u_\gamma) - u_\gamma^{(\bar\alpha_n)}$$
$$u_{\bar j}^{(\bar\alpha_j)} = u_{\bar j+1} \quad 1 \le \bar j \le \gamma - 1$$
$$u_\gamma^{(\bar\alpha_n)} = z_{M_n}^{(\bar\alpha_n)} + \mathscr{L}_s(z_{S_1}, z_{S_2}, \dots, z_{S_n}, u_1, u_2, \dots, u_\gamma) + K e_z$$

where $k = \begin{pmatrix} k_1 & k_2 & \cdots & k_n \end{pmatrix}$, $k_i > 0$ for $1 \le i \le n$ and $e_z = \begin{pmatrix} e_{z_1} & e_{z_2} & \cdots & e_{z_n} \end{pmatrix}^T$. Then, we have:

$$e_z^{(\bar\alpha)} = A e_z + \mathscr{G}_{z_M} \tag{9.4}$$

with

$$
A = \begin{pmatrix}
0 & 1 & 0 & \cdots & 0 & 0 \\
0 & 0 & 1 & \cdots & 0 & 0 \\
\vdots & & & \ddots & & \vdots \\
0 & 0 & 0 & \cdots & 1 & 0 \\
0 & 0 & 0 & \cdots & 0 & 1 \\
-k_1 & -k_2 & -k_3 & \cdots & -k_{n-1} & -k_n
\end{pmatrix}
$$

and

$$
\mathscr{G}_{zM} = \begin{pmatrix}
\mathscr{G}_1(z_{M_2}, z_{M_1}^{(\bar{\alpha}_1)}) \\
\mathscr{G}_2(z_{M_3}, z_{M_2}^{(\bar{\alpha}_2)}) \\
\vdots \\
\mathscr{G}_{n-2}(z_{M_{n-1}}, z_{M_{n-2}}^{(\bar{\alpha}_{n-2})}) \\
\mathscr{G}_{n-1}(z_{M_n}, z_{M_{n-1}}^{(\bar{\alpha}_{n-1})}) \\
0
\end{pmatrix}.
$$

In order to analyze the solution of (9.4), consider the following assumptions:

H1: Assume A is Hurwitz matrix with $\lambda_i(A) \neq \lambda_j(A)$.

H2: $\mathscr{G}_j$ is bounded, i.e., $\exists\, g_j \in \mathbb{R}^+$ such that

$$
|\mathscr{G}_j(z_{M_{j+1}}, z_{M_j}^{(\bar{\alpha}_j)})| \leq g_j
$$

for $1 \leq j \leq n-1$. $\qquad\qquad\square$

Remark 9.1 If the order of time-fractional derivatives coincides $\alpha = \bar{\alpha}$ (master and slave, respectively), the term $\mathscr{G}_j(z_{M_{j+1}}, z_{M_j}^{(\bar{\alpha}_j)})$ is zero for all j. Then, it is clear that the asymptotic stability of the zero solution of the Eq. (9.4) is directly obtained from Deng's Theorem [7] for all rational numbers $0 < \alpha = \bar{\alpha} < 1$.

Remark 9.2 Given the Hurwitz matrix $A \in \mathbb{R}^{n \times n}$ with different eigenvalues, i.e., $\lambda_i(A) \neq \lambda_j(A)$ there exists a linear transformation $V \in \mathbb{R}^{n \times n}$ such that $D = V^{-1}AV = \mathrm{diag}(\lambda_1, \ldots, \lambda_n)$, where matrix V is the Vandermonde matrix, and is given by:

$$
V := \begin{pmatrix}
1 & 1 & 1 & \cdots & 1 & 1 \\
\lambda_1 & \lambda_2 & \lambda_3 & \cdots & \lambda_{n-1} & \lambda_n \\
\vdots & \vdots & \vdots & & \vdots & \vdots \\
\lambda_1^{n-3} & \lambda_2^{n-3} & \lambda_3^{n-3} & \cdots & \lambda_{n-1}^{n-3} & \lambda_n^{n-3} \\
\lambda_1^{n-2} & \lambda_2^{n-2} & \lambda_3^{n-2} & \cdots & \lambda_{n-1}^{n-2} & \lambda_n^{n-2} \\
\lambda_1^{n-1} & \lambda_2^{n-1} & \lambda_3^{n-1} & \cdots & \lambda_{n-1}^{n-1} & \lambda_n^{n-1}
\end{pmatrix} \tag{9.5}
$$

with the associated characteristic monic polynomial $p(\lambda) = \prod_i^n (\lambda - \lambda_i) = p_0 + p_1 \lambda + \cdots + \lambda^n$. And, its inverse $V^{-1} = \Delta^{-T} V^T H$ where $\Delta = \mathrm{diag}(\dot{p}(\lambda_1), \ldots, \dot{p}(\lambda_n))$, $\dot{p}(\lambda_i)$ are evaluated derivatives of $p(\lambda)$ regarding to $\lambda = \lambda_i$ for $1 \le i \le n$ and H is the Hankel matrix given by:

$$H = \begin{pmatrix} p_1 & p_2 & p_3 & \cdots & p_{n-1} & 1 \\ p_2 & p_3 & p_4 & \cdots & 1 & 0 \\ \vdots & \vdots & \vdots & & \vdots & \vdots \\ p_{n-2} & p_{n-1} & 1 & \cdots & 0 & 0 \\ p_{n-1} & 1 & 0 & \cdots & 0 & 0 \\ 1 & 0 & 0 & \cdots & 0 & 0 \end{pmatrix}$$

where the coefficients of $p(\lambda)$ are $p_{i-1} = k_i$ for $1 \le i \le n$ (for further details of the inverse Vandermonde matrix see [8]).

Remark 9.3 Since the trajectories of a chaotic system are bounded, we can choose g_j as the $\sup_{1 \le j \le n-1} \mathscr{G}_j(\cdot)$ for each z_{M_j} with $1 \le j \le n-1$.

Assume V is invertible, taking the derivative of $\bar{e}_z = V^{-1} e_z$ in the trajectories of (9.4) we obtain:

$$\bar{e}_z^{(\bar{\alpha})} = D\bar{e}_z + \bar{\mathscr{G}}_{z_M} \tag{9.6}$$

where $\bar{\mathscr{G}}_{z_M} = V^{-1} \mathscr{G}_{z_M}$

$$\bar{\mathscr{G}}_{z_M} = \begin{pmatrix} \bar{\mathscr{G}}_1 \\ \bar{\mathscr{G}}_2 \\ \vdots \\ \bar{\mathscr{G}}_{n-2} \\ \bar{\mathscr{G}}_{n-1} \\ \bar{\mathscr{G}}_n \end{pmatrix}$$

with

$$\bar{\mathscr{G}}_i = \frac{\sum_{j=1}^{n-1} \left(\sum_{k=1}^{n-j} p_{j+k-1} \lambda_i^{k-1} + \lambda_i^{n-j} \right)}{\dot{p}(\lambda_i)} \mathscr{G}_j \tag{9.7}$$

for $1 \le i \le n$, $1 \le j \le n-1$. Note that for complex eigenvalues, λ_i should be replaced by its complex conjugate in (9.7). From assumption H2, we can establish the following

H2': $\bar{\mathscr{G}}_i$ is bounded, there exists $\bar{g}_i \in \mathbb{R}^+$ such that

$$|\bar{\mathscr{G}}_i| \le \bar{g}_i = \sum_{j=1}^{n-1} \left| \frac{\sum_{k=1}^{n-j} p_{j+k-1} \lambda_i^{k-1} + \lambda_i^{n-j}}{\dot{p}(\lambda_i)} \right| g_j, \quad 1 \le i \le n$$

Then, we obtain from (9.6) the following:

$$\bar{e}_{z_i}^{(\bar{\alpha}_i)} = \lambda_i \bar{e}_{z_i} + \bar{\mathscr{G}}_i, \quad \bar{e}_{z_i}(0) = \bar{e}_{z_i}(0), \quad 1 \leq i \leq n \tag{9.8}$$

Note that (9.8) is the initial value problem for a nonhomogeneous fractional differential equation under nonzero initial conditions with Caputo differential operator, due to $\bar{f}(\bar{e}_{z_i}, z_M) := \lambda_i \bar{e}_{z_i} + \bar{\mathscr{G}}_i(z_M)$ is a Lipschitz continuous function regarding to $\bar{e}_{z_i}$, that is to say, there exists a unique solution for (9.8) (see [9]). We can calculate the solution of (9.8) as follows:

$$\bar{e}_{z_i}(t) = \bar{e}_{z_{i0}} E_{\bar{\alpha}_i, i}(\lambda_i t^{\bar{\alpha}_i})$$
$$+ \int_0^t (t - \tau)^{\bar{\alpha}_i - 1} E_{\bar{\alpha}_i, \bar{\alpha}_i}(\lambda_i (t - \tau)^{\bar{\alpha}_i}) \bar{\mathscr{G}}_i(z_M(\tau)) d\tau$$

where $\bar{e}_{z_{i0}} = \bar{e}_{z_i}(0)$. Using Triangle and Cauchy–Schwarz inequalities, we can obtain the following:

$$|\bar{e}_{z_i}(t)| \leq |\bar{e}_{z_{i0}} E_{\bar{\alpha}_i, 1}(\lambda_i t^{\bar{\alpha}_i})|$$
$$+ \bar{g}_i \int_0^t |(t - \tau)^{\bar{\alpha}_i - 1} E_{\bar{\alpha}_i, \bar{\alpha}_i}(\lambda_i (t - \tau)^{\bar{\alpha}_i})| d\tau$$

Due to Property 2.1 (Chap. 2) of Mittag-Leffler functions and $\lambda_i < 0$ we have $(t - \tau)^{\bar{\alpha}_i - 1} E_{\bar{\alpha}_i, \bar{\alpha}_i}(\lambda_i (t - \tau)^{\bar{\alpha}_i})$ and $E_{\bar{\alpha}_i, 1}(\lambda_i, t)^{\bar{\alpha}_i}$ are not negative, then

$$|\bar{e}_{z_i}(t)| \leq |\bar{e}_{z_{i0}}| E_{\bar{\alpha}_i, 1}(\lambda_i t^{\bar{\alpha}_i}) \tag{9.9}$$
$$+ \bar{g}_i \int_0^t (t - \tau)^{\bar{\alpha}_i - 1} E_{\bar{\alpha}_i, \bar{\alpha}_i}(\lambda_i (t - \tau)^{\bar{\alpha}_i}) d\tau$$

and from Property 2.2 (Chap. 2)

$$|\bar{e}_{z_i}(t)| \leq |\bar{e}_{z_{i0}}| E_{\bar{\alpha}_i, 1}(\lambda_i t^{\bar{\alpha}_i}) + \bar{g}_i t^{(\bar{\alpha}_i)} E_{\bar{\alpha}_i, \bar{\alpha}_i + 1}(\lambda_i t^{\bar{\alpha}_i})$$

Note that, if $t \to \infty$, from (9.9) with $\mu = 3\pi \dfrac{\bar{\alpha}_i}{4}$ using Theorem 2.1 (Chap. 2)

$$\lim_{t \to \infty} |\bar{e}_{z_i}(t)| \leq |\bar{e}_{z_{i0}}| \lim_{t \to \infty} E_{\bar{\alpha}_i, 1}(\lambda_i t^{\bar{\alpha}_i})$$
$$+ \bar{g}_i \lim_{t \to \infty} t^{\bar{\alpha}_i} E_{\bar{\alpha}_i, \bar{\alpha}_i + 1}(\lambda_i t^{\bar{\alpha}_i}) = \dfrac{\bar{g}_i}{|\lambda_i|}$$

Then, system (9.8) converges asymptotically to a compact set $\left\{\, \bar{e}_{z_i} \mid |\bar{e}_{z_i}| \leq \bar{g}_i/|\lambda_i| \,\right\}$. The former enable us to obtain a bound for the error vector in transformed coordinates. Assume $e_z = V\bar{e}_z$, then:

$$e_{z_i} = \sum_{l=1}^{n} \lambda_l^{i-1} \bar{e}_{z_l} \quad 1 \leq i \leq n$$

using Triangle inequality and the limit when $t \to \infty$, we have the following:

$$\lim_{t\to\infty} |e_{z_i}| \leq \bar{r}_i := \sum_{l=1}^{n} |\lambda_l|^{i-1} \lim_{t\to\infty} |\bar{e}_{z_l}|$$

Finally, the next estimate is fulfilled[1]

$$\lim_{t\to\infty} \|e_z\|_\infty \leq r$$

Then the solution of (9.4) converges asymptotically to a compact set $\bar{B}_r(0)$ with $r = \max_{1 \leq i \leq n} \bar{r}_i$.

Remark 9.4 It is not hard to see that when $\alpha \neq \bar{\alpha}$:

$$\lim_{t\to\infty} \| (z_M, z_S) \|_{M_z} \leq \sqrt{\frac{n}{2}} \lim_{t\to\infty} \|e_z\|_\infty \leq \sqrt{\frac{n}{2}} r$$

and when $\alpha = \bar{\alpha}$:

$$\lim_{t\to\infty} \| (z_M, z_S) \|_{M_z} = 0$$

where

$$\| (z_M, z_S) \|_{M_z} := \inf_{(\hat{z}_M, \hat{z}_S) \in M_z} \left\{ \| (z_M, z_S) - (\hat{z}_M, \hat{z}_S) \| \right\},$$

with

$$M_z := \{(z_M, z_S) \mid z_M = z_S\}.$$

From Lemma 9.1 and Theorem 9.1 we can establish the following result.

---

[1] For $x = (x_1, x_2, \ldots, x_n)^T \in \mathbb{R}^n$, $\| x \|_\infty := \max\{|x_1|, |x_2|, \ldots, |x_n|\}$.

Corollary 9.1 *All incommensurate fractional order systems are in a state of FGS if and only if are PV.*

Proof The proof is immediate and is omitted.   □

9.3   Numerical Results

Consider the incommensurate fractional order Chua–Hartley system [10] as the master system:

$$
\begin{aligned}
x_1^{(\alpha_1)} &= ax_2 + \frac{ax_1}{7} - \frac{2ax_1^3}{7} \\
x_2^{(\alpha_2)} &= x_1 - x_2 + x_3 \\
x_3^{(\alpha_3)} &= -\beta x_2 \\
y &= x_1
\end{aligned}
\tag{9.10}
$$

where $\alpha = (0.94\ 0.98\ 0.92)$, with $a = 12.75$, $\beta = 100/7$, the fractional incommensurate differential primitive element is chosen as the output of system $y_M = x_1$.
Then the corresponding coordinate transformation is given by:

$$
\begin{pmatrix} z_{1_M} \\ z_{2_M} \\ z_{3_M} \end{pmatrix} =
\begin{pmatrix} y_M \\ y_M^{(\alpha_1)} \\ y_M^{(\alpha_2)} \end{pmatrix} = \phi_M(x_M)
\tag{9.11}
$$

$$
= \begin{pmatrix}
x_{1_M} \\
ax_{2_M} + \dfrac{ax_{1_M}}{7} - \dfrac{2ax_{1_M}^3}{7} \\
a(x_{1_M} - x_{2_M} + x_{3_M}) + \dfrac{ax_{1_M}^{(\alpha_2)}}{7} - \dfrac{2a(x_{1_M})^{3(\alpha_2)}}{7}
\end{pmatrix}
$$

From (9.11), we obtain the IFGOCF.

$$
z^{(\alpha)}{}_M =
\begin{pmatrix} z^{(\alpha_1)}{}_{1_M} \\ z^{(\alpha_2)}{}_{2_M} \\ z^{(\alpha_3)}{}_{3_M} \end{pmatrix} =
\begin{pmatrix} z_{2_M} \\ z_{3_M} \\ \psi_M(x_M) \end{pmatrix}
\tag{9.12}
$$

$$
= \begin{pmatrix}
ax_{2_M} + \dfrac{ax_{1_M}}{7} - \dfrac{2ax_{1_M}^3}{7} \\
a(x_{1_M} - x_{2_M} + x_{3_M}) + \dfrac{ax_{1_M}^{(\alpha_2)}}{7} - \dfrac{2a(x_{1_M})^{3(\alpha_2)}}{7} \\
ax_{1_M}^{(\alpha_3)} - ax_{2_M}^{(\alpha_3)} - a\beta x_{2_M} + D^{\alpha_2}D^{\alpha_1}\left(\dfrac{ax_{1_M}}{7}\right) - D^{\alpha_3}D^{\alpha_2}\left(\dfrac{2a(x_{1_M})^3}{7}\right)
\end{pmatrix}
$$

Let the slave system be the incommensurate fractional order chaotic Rössler system given by [11]:

$$
\begin{aligned}
x_1^{(\bar{\alpha}_1)} &= -x_2 - x_3 \\
x_2^{(\bar{\alpha}_2)} &= x_1 + 0.63x_2 \\
x_3^{(\bar{\alpha}_3)} &= 0.2 + x_3(x_1 - 10) \\
y &= x_2
\end{aligned}
\tag{9.13}
$$

where $\bar{\alpha} = (0.9\ 0.8\ 0.7)$, the fractional incommensurate differential primitive element is chosen as the output of system (9.13) $y_S = x_2 + u_1$.

We propose the coordinate transformation:

$$
\begin{pmatrix} z_{1_S} \\ z_{2_S} \\ z_{3_S} \end{pmatrix} = \begin{pmatrix} y_S \\ y_S^{(\bar{\alpha}_2)} \\ y_S^{(\bar{\alpha}_1)} \end{pmatrix} = \phi_S(x_S)
\tag{9.14}
$$

$$
= \begin{pmatrix} x_{2_S} + u_1 \\ x_{1_S} + 0.63x_{2_S} + u_2 \\ -x_{2_S} - x_{3_S} + 0.63x_{2_S}^{(\bar{\alpha}_1)} + u_3 \end{pmatrix}
$$

Then, the IFGOCF of system (9.13) is given by:

$$
z^{(\bar{\alpha})}{}_S = \begin{pmatrix} z^{(\bar{\alpha}_2)}{}_{1_S} \\ z^{(\bar{\alpha}_1)}{}_{2_S} \\ z^{(\bar{\alpha}_3)}{}_{3_S} \end{pmatrix} = \begin{pmatrix} z_{2_S} \\ z_{3_S} \\ \psi_S(x_S) + u_3^{(\bar{\alpha}_3)} \end{pmatrix}
\tag{9.15}
$$

$$
= \begin{pmatrix} x_{1_S} + 0.63x_{2_S} + u_2 \\ -x_{2_S} - x_{3_S} + 0.63x_{2_S}^{(\bar{\alpha}_1)} + u_3 \\ -x_{2_S}^{(\bar{\alpha}_3)} - 0.2 - x_{3_S}(x_{1_S} - 10) + D^{\bar{\alpha}_3}D^{\bar{\alpha}_1}0.63x_{2_S} + u_3^{(\bar{\alpha}_3)} \end{pmatrix}
$$

According to methodology proposed, consider the synchronization error as:

$$
e_i = z_{i_M} - z_{i_S}, \quad 1 \le i \le 3
\tag{9.16}
$$

i.e.,

$$
\begin{aligned}
e_1 &= z_{1_M} - z_{1_S} \\
e_2 &= z_{2_M} - z_{2_S} \\
e_3 &= z_{3_M} - z_{3_S}
\end{aligned}
$$

Taking the Caputo derivative $\bar{\alpha}_i$ of (9.16), we have:

$$_0D_t^{\bar{\alpha}_i}e_i = e_i^{(\bar{\alpha}_i)} = z_{i_M}^{(\bar{\alpha}_i)} - z_{i_S}^{(\bar{\alpha}_i)}$$

for $\bar{\alpha}_i$ in the order that the transformate coordinates appear in slave system. The incommensurate error dynamics is the following:

$$e_1^{(\bar{\alpha}_2)} = e_2 + \mathcal{G}_1$$
$$e_2^{(\bar{\alpha}_1)} = e_3 + \mathcal{G}_2$$
$$e_3^{(\bar{\alpha}_3)} = z_{3_M}^{(\bar{\alpha}_3)} - \Psi_S - u_3^{(\bar{\alpha}_3)}$$

where $\mathcal{G}_1 := z_{1_M}^{(\bar{\alpha}_2)} - z_{2_M}$ and $\mathcal{G}_2 := z_{2_M}^{(\bar{\alpha}_1)} - z_{3_M}$.

Taking $u_3^{(\bar{\alpha}_3)} = z_{3_M}^{(\bar{\alpha}_3)} - \Psi_S + ke$, where $k = (k_1 \; k_2 \; k_3)$, $k_1, k_2, k_3 > 0$ and $e = (e_1 \; e_2 \; e_3)^T$, then:

$$\begin{pmatrix} e_1^{(\bar{\alpha}_2)} \\ e_2^{(\bar{\alpha}_1)} \\ e_3^{(\bar{\alpha}_3)} \end{pmatrix} = \begin{pmatrix} 0 & 1 & 0 \\ 0 & 0 & 1 \\ -k_1 & -k_2 & -k_3 \end{pmatrix} \begin{pmatrix} e_1 \\ e_2 \\ e_3 \end{pmatrix} + \begin{pmatrix} \mathcal{G}_1 \\ \mathcal{G}_2 \\ 0 \end{pmatrix} \qquad (9.17)$$

Choosing the gains $k = (170, 476, 150)$ with initial conditions $z_{M0} = (6 \; 1 \; {-6})$ and $z_{S0} = (5 \; 1 \; {-6})$. We can numerically estimate the bounds for $\mathcal{G}_1$ and $\mathcal{G}_2$ as $g_1 = 0.4994$ and $g_2 = 1.1642$ respectively. Due to $\lambda_1 = -146.764$, $\lambda_2 = -0.41$, $\lambda_3 = -2.825$ are distinct eigenvalues and the corresponding matrix is:

$$V = \begin{pmatrix} 1 & 1 & 1 \\ \lambda_1 & \lambda_2 & \lambda_3 \\ \lambda_1^2 & \lambda_2^2 & \lambda_3^2 \end{pmatrix}$$

is nonsingular, (9.17) can be expressed as follows:

$$\begin{pmatrix} \bar{e}_1^{(\bar{\alpha}_2)} \\ \bar{e}_2^{(\bar{\alpha}_1)} \\ \bar{e}_3^{(\bar{\alpha}_3)} \end{pmatrix} = \begin{pmatrix} \lambda_1 & 0 & 0 \\ 0 & \lambda_2 & 0 \\ 0 & 0 & \lambda_3 \end{pmatrix} \begin{pmatrix} \bar{e}_1 \\ \bar{e}_2 \\ \bar{e}_3 \end{pmatrix} + \begin{pmatrix} \bar{\mathcal{G}}_1 \\ \bar{\mathcal{G}}_2 \\ \bar{\mathcal{G}}_3 \end{pmatrix},$$

where

$$\bar{\mathcal{G}}_i = \frac{(k_2 + k_3\lambda_i + \lambda_i^2)\mathcal{G}_1 + (k_3 + \lambda_3)\mathcal{G}_2}{k_2 + 2k_3\lambda_i + 3\lambda_i^2}, \quad 1 \le i \le 3 \qquad (9.18)$$

Consequently, we can calculate the bounds for $\mathscr{G}_1$, $\mathscr{G}_2$, and $\mathscr{G}_3$ as $\bar{g}_1 = 0.0002$, $\bar{g}_2 = 1.0784$ and $\bar{g}_3 = 0.5792$ respectively. Finally, we conclude that error e_i is bounded for all i in an compact set $\bar{B}_r(0)$, with $r = 2.8356$.

The original variables can be obtained from the inverse transformation, since they satisfy the IFAO condition, that is to say:

$$\begin{pmatrix} x_{1_M} \\ x_{2_M} \\ x_{3_M} \end{pmatrix} = \begin{pmatrix} z_{1_M} \\ \dfrac{z_{2_M}}{a} - \dfrac{z_{1_M}}{7} + \dfrac{2z_{1_M}^3}{7} \\ \dfrac{z_{3_M}}{a} - z_{1_M} + \dfrac{z_{2_M}}{a} - \dfrac{z_{1_M}}{7} + \dfrac{2z_{1_M}^3}{7} - \dfrac{z_{1_M}^{(\alpha_2)}}{7} + \dfrac{2z_{1_M}^{3(\alpha_2)}}{7} \end{pmatrix}$$

$$= \phi_M^{-1}(z_M)$$

$$\begin{pmatrix} x_{1_S} \\ x_{2_S} \\ x_{3_S} \end{pmatrix} = \begin{pmatrix} z_{2_S} - 0.63z_{1_S} + 0.63u_1 - u_2 \\ z_{1_S} + u_1 \\ -z_{3_S} - z_{1_S} + u_1 + 0.63(z_1 - u_1)^{(\bar{\alpha}_1)} + u_3 \end{pmatrix}$$

$$= \phi_S^{-1}(z_S)$$

In Figs. 9.1 and 9.2, the transformed trajectories are displayed. In Figs. 9.3 and 9.4, the original reconstructed trajectories of fractional order systems are shown. Finally, the error in transformed and original coordinates is shown in Figs. 9.5 and 9.6.

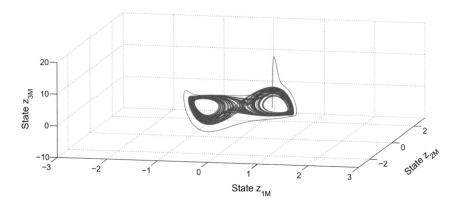

Fig. 9.1 Master system in transformate coordinates

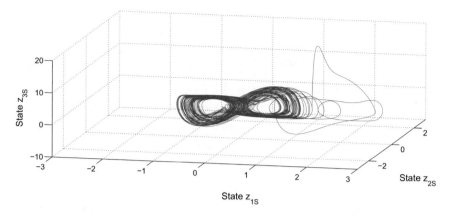

Fig. 9.2 Slave system in transformate coordinates

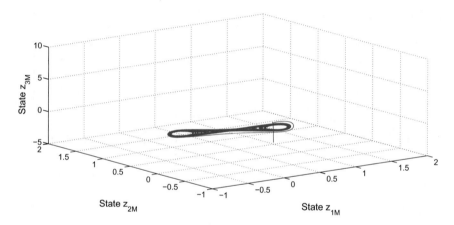

Fig. 9.3 Master system in original coordinates

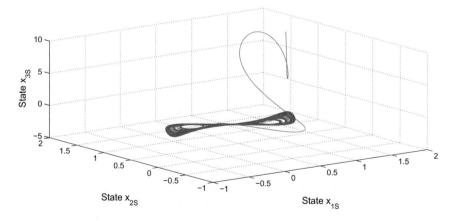

Fig. 9.4 Slave system in original coordinates

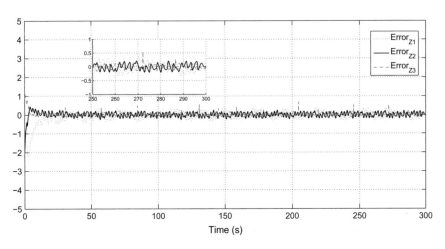

Fig. 9.5 Synchronization errors in Transformate coordinates

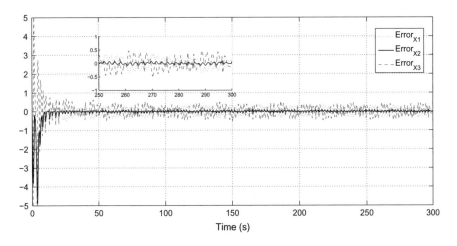

Fig. 9.6 Synchronization errors in original coordinates

9.4   Concluding Remarks

In this chapter, we have proposed an incommensurate fractional order dynamical controller for the Generalized Synchronization problem of incommensurate fractional order systems, where an Incommensurate Fractional Observability Canonical Form for these class of systems is given. We have introduced a new property called fractional algebraic observability for Incommensurate Fractional Order Systems (IFAO). As far as we know, the perfect synchronization has not been achieved in IFOCS. Therefore, we are talking about quasi-synchronization for IFOCS. The method was applied to an incommensurate fractional order Chua–Hartley system and Rössler system and we can observe that the synchronization error is contained in a compact region and numerical results show the effectiveness of the methodology proposed.

References

1. N.F. Rulkov, M.M. Sushchik, L.S. Tsimring, H. Abarbanel: Generalized synchronization of chaos in directionally coupled chaotic systems. Phys. Rev. E. 51, 980–994 (1995).
2. R. Martínez-Guerra, J. Mata-Machuca: Fractional generalized synchronization in a class of nonlinear fractional order system, Nonlinear Dynamics, Vol. 77, 1237–1244 (2014).
3. A. Razminia, V.J. Majd, D. Baleanu: Chaotic incommensurate fractional order Rössler system: active control and synchronization, Advances in Difference Equations, 1–12 (2011).
4. S.S. Delshad, M.M. Asheghan, M.H. Beheshti: Synchronization of N-coupled incommensurate fractional-order chaotic systems with ring connection, Communications in Nonlinear Science and Numerical Simulation, 3815–3824 (2011).
5. J.W. Wang, Y.B Zhang: Synchronization in coupled nonidentical incommensurate fractional-order systems, Physics Letters A, 202–207 (2009).
6. A. Boulkroune, A. Bouzeriba, T. Bouden: Fuzzy generalized projective synchronization of incommensurate fractional-order chaotic systems, Neurocomputing, Vol. 173, 606–614 (2016).
7. Deng, Weihua, Changpin Li, and Jinhu L, Stability analysis of linear fractional differential system with multiple time delays, Nonlinear Dynamics 48(4), 409–416, (2007).
8. U. Luther and K. Rost: Matrix exponentials and inversion of confluent Vandermonde matrices. *Electronic Transactions on Numerical Analysis*, 18, 9–100 (2004).
9. A. A. Kilbas, H. M. Srivastava and J. J. Trujillo: Theory and Applications of Fractional Differential Equations, Elsevier, 1st ed., (2006).
10. R. Caponetto, G. Dongola, L. Fortuna and I. Petrás: Fractional Order Systems, World Scientific Series on Nonlinear Science, Series A. Vol 72, (2010).
11. M. S. Tavazoei, M. Haeri: Chaotic attractors in incommensurate fractional order systems, Physica D Nonlinear Phenomena, 2628–2637 (2008).

Chapter 10
Synchronization and Anti-synchronization of Fractional Order Chaotic Systems by Means of a Fractional Integral Observer

10.1 Introduction

The problem of anti-synchronization is another phenomenon of interest that occurs in chaotic oscillators. This problem has appeared in modern repetitions of Huygens' experiments [1], lasers [2, 3], saltwater oscillators [4], and some biological systems where a nonchaotic signal is generated [5]. Anti-synchronization has been treated as a direct modification of synchronization, simply with a sign change in the condition required for the error, and has been attacked with methods such as the active control [6, 7] and the sliding mode control [8]. It can also be induced by noise [9].

Recently, systems with fractional dynamics, i.e., systems whose mathematical model is represented with derivatives and integrals of non-integer-order, have been of great interest. This is mainly given to their applications to interdisciplinary areas such as material science [10], electromagnetism [11], electromechanics [12], and thermal systems [13]; in addition, in certain cases, fractional equations give better approximations of the behavior of the systems than the integer ones. Furthermore, it has been found that these systems can present chaotic behavior, and also some strategies to control or synchronize them have been developed [14–18].

This chapter deals with the synchronization and anti-synchronization problems in the Lorenz chaotic system with commensurate fractional dynamics, i.e., where the fractional order of the dynamics of all the states is the same, and in the Rössler chaotic system with incommensurate fractional dynamics. A reduced-order fractional integral observer is proposed, whose design is based on the fractional algebraic observability property, in order to estimate the states of the master system and build the slave system that will synchronize with it. This observer is also used to estimate some fractional derivatives of the output that appears in the slave system dynamics. After applying the methodology, simulations are performed in order to obtain numerical results.

The chapter is divided as follows. In Sect. 10.2, some concepts and results from fractional calculus are introduced. In Sect. 10.3, the reduced-order fractional

© Springer Nature Switzerland AG 2018
R. Martínez-Guerra and C. A. Pérez-Pinacho, *Advances in Synchronization of Coupled Fractional Order Systems*, Understanding Complex Systems, https://doi.org/10.1007/978-3-319-93946-9_10

integral observer to be used is proposed, the synchronization and anti-synchronization problems are defined and the proposed methodology for their solution by means of the observer is presented. In Sect. 10.2, the methodology is applied to the fractional Lorenz and Rössler systems and numerical simulations are performed. Finally, Sect. 10.3 concludes the results of the chapter.

10.1.1   Reduced-Order Fractional Integral Observer

Once determining that all the unknown states of the system are observable, the fractional synchronization problem can be solved using the master–slave configuration by means of an observer. Consider again system (2.21) without u:

$$D^\alpha x(t) = f(x)$$
$$y(t) = h(x).$$

This dynamics can be extended to include the unknown vector state in a new variable η with unknown dynamics:

$$D^\alpha x = f(x, \eta)$$
$$D^\alpha \eta = \Delta(x)$$
$$y = h(x).$$

The problem is to construct the state η in order to determine the value of the desired state. But, given that the model has an unknown part, a fractional classical Luenberguer observer cannot be constructed. So, for this matter, the following reduced-order fractional integral observer (ROFIO) is proposed:

$$D^\alpha \hat{\eta}_i = K_{i0}(\eta_i - \hat{\eta}_i) + K_{i1} I^\alpha (\eta_i - \hat{\eta}_i). \tag{10.1}$$

This observer considers a proportional corrective term of the estimation error, followed by a fractional integral term of the error in order to improve its convergence.

Remark 10.1 The ROFIO is model-free, i.e., it does not require to know the dynamics of the states, using only the FAO (or IFAO) condition to reconstruct them. Thus, it has an advantage against other methods and observers used for synchronization that require full knowledge of the system dynamics.

In order to work with the ROFIO, it is assumed that the following hypotheses are satisfied:

H1: η_i satisfies the FAO (or IFAO) condition.
H2:   Let an auxiliary variable γ_i be a C^1 real-valued function.
H3: $\Delta_i(x)$ is bounded, i.e., $\exists\, N_i \in \mathbb{R}^+$ such that $\|\Delta_i(x)\| \leq N_i, \forall x \in \Omega \subset \mathbb{R}^n$.

Theorem 10.1 *The reduced-order fractional integral observer (10.1) is Mittag-Leffler stable.*

Proof Define the observer error as $e_i = \eta_i - \hat{\eta}_i$. Consider the following Lyapunov function candidate:

$$V(e_i) = e_i^2. \tag{10.2}$$

Note that $V(e_i)$ satisfies the first inequality of Theorem 2.4 since

$$\alpha_1 \|e_i\| \leq V(e_i) \leq \alpha_2 \|e_i\| \tag{10.3}$$

with $a = b = \alpha_1 = 1$ and $\alpha_2 = \sup(\|e_i\|)$.

By Lemma 2.3, it follows that

$$D^\alpha V(e_i) = D^\alpha e_i^2 \leq 2e_i D^\alpha e_i.$$

Therefore,

$$\begin{aligned}
D^\alpha V(e_i) &\leq 2e_i D^\alpha e_i = 2e_i D^\alpha (\eta_i - \hat{\eta}_i) \\
&= 2e_i(\Delta_i - K_{i0}e_i - K_{i1}I^\alpha e_i) \\
&\leq 2e_i \Delta_i - 2K_{i1}e_i I^\alpha e_i \\
&\leq 2N_i \|e_i\| - 2K_{i1}\|e_i\|\|I^\alpha e_i\| \\
&\leq -(2K_{i1}|I^\alpha e_i| - 2N_i)\|e_i\|.
\end{aligned}$$

Then

$$D^\alpha V(e_i) \leq -\alpha_3 \|e_i\| \tag{10.4}$$

with $\alpha_3 = 2K_{i1}|I^\alpha e_i| - 2N_i$.

Therefore, if $\alpha_3 > 0$, from Theorem 2.4 and Eqs. (10.2)–(10.4), it is concluded that the origin of system (10.1) is Mittag-Leffler stable. □

10.1.2 Fractional Synchronization Problem

In this chapter, the fractional synchronization problem is solved as follows. The original fractional chaotic system will be known as the "master", because it acts as a driving system by means of its measurable output. Then, the ROFIO will use this output in order to synchronize the dynamics of the estimated states with those of the master system; hence, the ROFIO is known as the "slave" system. This problem is usually determined by the analysis of the fractional synchronization error dynamics, which is defined as follows:

$$e_i = \eta_i - \hat{\eta}_i. \tag{10.5}$$

Considering the fractional dynamics of this error, the following is obtained:

$$D^{\alpha} e_i = D^{\alpha} \eta_i - D^{\alpha} \hat{\eta}_i$$
$$= D^{\alpha} \eta_i - K_{i0} e_i - K_{i1} I^{\alpha} e_i$$
$$D^{\alpha} e_i + K_{i0} e_i + K_{i1} I^{\alpha} e_i = D^{\alpha} \eta_i.$$

The last equation is transformed to the Laplace domain:

$$s^{\alpha} E_i(s) - s^{\alpha-1} e_i(0) + K_{i0} E_i(s) + K_{i1} s^{-\alpha} E_i(s) = s^{\alpha} H_i(s) - s^{\alpha-1} \eta_i(0),$$

from where the following solution is obtained:

$$E_i(s) = \frac{s^{\alpha} H_i(s) - s^{\alpha-1} \eta_i(0) + s^{\alpha-1} e_i(0)}{s^{\alpha} + K_{i0} + K_{i1} s^{-\alpha}}$$
$$= \frac{s^{\alpha} (s^{\alpha} H_i(s) - s^{\alpha-1} \eta_i(0) + s^{\alpha-1} e_i(0))}{s^{2\alpha} + K_{i0} s^{\alpha} + K_{i1}}.$$

Note that this equation can be rewritten as

$$E_i(s) = \frac{s^{\alpha} (s^{\alpha} H_i(s) - s^{\alpha-1} \eta_i(0) + s^{\alpha-1} e_i(0))}{(s^{\alpha} + \lambda_1)(s^{\alpha} + \lambda_2)} = \frac{l_1}{s^{\alpha} + \lambda_1} + \frac{l_2}{s^{\alpha} + \lambda_2},$$

with $l_1, l_2 \in \mathbb{R}$, and this solution in time domain reads as follows:

$$e_i = l_1 t^{\alpha-1} \mathscr{E}_{\alpha,\alpha}(-\lambda_1 t^{\alpha}) + l_2 t^{\alpha-1} \mathscr{E}_{\alpha,\alpha}(-\lambda_2 t^{\alpha}).$$

So,

$$\|e_i\| \le \|t^{\alpha-1}\| (|l_1| \|\mathscr{E}_{\alpha,\alpha}(-\lambda_1 t^{\alpha})\| + |l_2| \|\mathscr{E}_{\alpha,\alpha}(-\lambda_2 t^{\alpha})\|),$$

and hence it can be observed that the norm of the estimation error is bounded by the absolute values of the scalars l_1 and l_2 and by the norms of $t^{\alpha-1}$ and the Mittag-Leffler functions $\mathscr{E}_{\alpha,\alpha}(-\lambda_1 t^{\alpha})$ and $\mathscr{E}_{\alpha,\alpha}(-\lambda_2 t^{\alpha})$. Given that $0 < \alpha < 1, \alpha - 1 < 0$, and thus the function $t^{\alpha-1}$ decreases. Furthermore, if gains K_{i0} and K_{i1} are chosen such that the polynomial $s^{2\alpha} + K_{i0} s^{\alpha} + K_{i1} = (s^{\alpha} + \lambda_1)(s^{\alpha} + \lambda_2)$ is stable, i.e., such that the following condition is fulfilled [19]:

$$|arg(\lambda_i)| > \alpha \frac{\pi}{2}, \quad i = 1, 2,$$

then the polynomial is stable, and the Mittag-Leffler functions $\mathscr{E}_{\alpha,\alpha}(-\lambda_1 t^{\alpha})$ and $\mathscr{E}_{\alpha,\alpha}(-\lambda_2 t^{\alpha})$ tend to the origin of the error dynamics, which makes $\hat{\eta}_i$ to follow η_i. Thus, the fractional synchronization problem is solved. $\qquad \square$

10.1.3 Fractional Anti-synchronization Problem

The problem of fractional anti-synchronization consists in obtaining also an estimation of the unknown states of the master system, but instead of having the state of the slave system converges to the value of the state of the master, it will converge to that value with opposite sign. For this, a new variable $\hat{\xi}_i = -\hat{\eta}_i$ is introduced, which is the estimated state of the anti-synchronized system. So, the fractional anti-synchronization error dynamics is defined as follows:

$$\bar{e}_i = \eta_i + \hat{\xi}_i. \tag{10.6}$$

In a procedure similar to the former, the following equation is obtained:

$$\|\bar{e}_i\| \le \|t^{\alpha-1}\| (|l_1| \|\mathscr{E}_{\alpha,\alpha}(-\lambda_1 t^\alpha)\| + |l_2| \|\mathscr{E}_{\alpha,\alpha}(-\lambda_2 t^\alpha)\|),$$

hence selecting the appropriate values of gains K_{i0} and K_{i1}, the polynomial $s^{2\alpha} + K_{i0}s^\alpha + K_{i1} = (s^\alpha + \lambda_1)(s^\alpha + \lambda_2)$ is stable, and the Mittag-Leffler functions $\mathscr{E}_{\alpha,\alpha}(-\lambda_1 t^\alpha)$ and $\mathscr{E}_{\alpha,\alpha}(-\lambda_2 t^\alpha)$ tend to the origin of the error dynamics, which makes $\hat{\xi}_i$ to follow η_i but with an inverse sign. Thus, the fractional anti-synchronization problem is solved.

Remark 10.2 This methodology works for commensurate and incommensurate-order systems, given that the respective unknown variables satisfy the FAO (or IFAO) property.

10.2 Application to Fractional Chaotic Systems

In this section, the ROFIO is used to achieve synchronization and anti-synchronization in two fractional chaotic systems, namely, the Lorenz and the Rössler oscillators. The former will be treated as a commensurate-order system, and the latter as an incommensurate-order one.

10.2.1 Fractional Lorenz System

The Lorenz oscillator is an attractor named after Edward M. Lorenz, who derived it from the simplified equations of turbulent convection rolls arising in the equations of the atmosphere [20]. This system is related to the so-called "butterfly effect". The generalization to fractional dynamics was presented in [21].

Consider the commensurate fractional order chaotic Lorenz system:

$$D^\alpha x_1 = \sigma(x_2 - x_1) \tag{10.7}$$
$$D^\alpha x_2 = \rho x_1 - x_2 - x_1 x_3$$
$$D^\alpha x_3 = x_1 x_2 - \beta x_3$$
$$y = x_1$$

with $\sigma, \rho, \beta > 0$.

First, it has to be verified that the unknown states, x_2 and x_3, satisfy the FAO condition. From (10.7), the following equations can be obtained:

$$x_2 = \phi(y, D^\alpha y) = (1/\sigma)(\sigma y + D^\alpha y) \tag{10.8}$$
$$x_3 = \phi(y, D^\alpha y, D^{2\alpha} y) = \rho - 1 - (1/\sigma y)(D^\alpha y(\sigma + 1) + D^{2\alpha} y),$$

and thus, both states are algebraically observable, so the ROFIO can be built.

Let $\eta_2 = x_2$. Using (10.8), the ROFIO for this variable is

$$D^\alpha \hat{\eta}_2 = K_{10}(\eta_2 - \hat{\eta}_2) + K_{11}I^\alpha(\eta_2 - \hat{\eta}_2)$$
$$= K_{10}\left(\frac{1}{\sigma}(\sigma y + D^\alpha y) - \hat{\eta}_2\right) + K_{11}I^\alpha\left(\frac{1}{\sigma}(\sigma y + D^\alpha y) - \hat{\eta}_2\right).$$

Define the auxiliary variable $\gamma_1 = \hat{\eta}_2 - \frac{K_{10}}{\sigma}y$. Then,

$$D^\alpha \gamma_1 = D^\alpha \hat{\eta}_2 - \frac{K_{10}}{\sigma}D^\alpha y$$
$$= K_{10}\left(y - \hat{\eta}_2\right) + \frac{K_{11}}{\sigma}y + K_{11}I^\alpha\left(y - \hat{\eta}_2\right).$$

So, the ROFIO for $\hat{\eta}_2$ is

$$D^\alpha \gamma_1 = K_{10}\left(y - \gamma_1 - \frac{K_{10}}{\sigma}y\right) + \frac{K_{11}}{\sigma}y + K_{11}I^\alpha\left(y - \gamma_1 - \frac{K_{10}}{\sigma}y\right) \tag{10.9}$$
$$\hat{\eta}_2 = \gamma_1 + \frac{K_{10}}{\sigma}y. \tag{10.10}$$

Now let $\eta_3 = x_3$. Note that from (10.7), the following relation can be obtained:

$$x_3 = (1/\beta)[(y/\sigma)(\sigma y + D^\alpha y) - D^\alpha \eta_3]. \tag{10.11}$$

Using this equation, the ROFIO for this variable is

$$D^\alpha \hat{\eta}_3 = K_{20}(\eta_3 - \hat{\eta}_3) + K_{21}I^\alpha(\eta_3 - \hat{\eta}_3)$$

$$= K_{20}\left(\frac{1}{\beta}\left(\frac{1}{\sigma}y(\sigma y + D^\alpha y) - D^\alpha \hat{\eta}_3\right) - \hat{\eta}_3\right)$$

$$+ K_{21}I^\alpha\left(\frac{1}{\beta}\left(\frac{1}{\sigma}y(\sigma y + D^\alpha y) - D^\alpha \hat{\eta}_3\right) - \hat{\eta}_3\right).$$

After some algebraic manipulations, the following is obtained:

$$D^\alpha \hat{\eta}_3 = \frac{K_{20}}{(\beta + K_{20})\sigma}yD^\alpha y$$

$$+ \frac{K_{20}\beta}{\beta + K_{20}}\left[\left(\frac{1}{\beta} - \frac{K_{20}}{2(\beta + K_{20})\sigma}\right)y^2 - \hat{\eta}_3 + \frac{K_{20}}{2(\beta + K_{20})\sigma}y^2\right]$$

$$- \frac{K_{20}K_{21}}{\beta + K_{20}}\hat{\eta}_3 + \frac{K_{20}K_{21}\beta}{\beta + K_{20}}I^\alpha\left(\frac{1}{\beta\sigma}y(\sigma y + D^\alpha y) - \hat{\eta}_3\right).$$

Define the auxiliary variable $\gamma_2 = \hat{\eta}_3 - \frac{K_{20}}{2(\beta+K_{20})\sigma}y^2$. Then,

$$D^\alpha \gamma_2 = D^\alpha \hat{\eta}_3 - \frac{K_{20}}{(\beta + K_{20})\sigma}yD^\alpha y$$

$$= \frac{K_{20}\beta}{\beta + K_{20}}\left[\left(\frac{1}{\beta} - \frac{K_{20}}{2(\beta + K_{20})\sigma}\right)y^2 - \gamma_2\right]$$

$$- \frac{K_{20}K_{21}}{\beta + K_{20}}\hat{\eta}_3 + \frac{K_{20}K_{21}\beta}{\beta + K_{20}}I^\alpha\left(\frac{1}{\beta\sigma}y(\sigma y + D^\alpha y) - \hat{\eta}_3\right).$$

So, the ROFIO for $\hat{\eta}_3$ is

$$D^\alpha \gamma_2 = \frac{K_{20}\beta}{\beta + K_{20}}\left[\left(\frac{1}{\beta} - \frac{K_{20}}{2(\beta + K_{20})\sigma}\right)y^2 - \gamma_2\right] \tag{10.12}$$

$$- \frac{K_{20}K_{21}}{\beta + K_{20}}\left(\gamma_2 + \frac{K_{20}}{2(\beta + K_{20})\sigma}y^2\right)$$

$$+ \frac{K_{20}K_{21}\beta}{\beta + K_{20}}I^\alpha\left(\frac{1}{\beta\sigma}y(\sigma y + D^\alpha y) - \gamma_2 - \frac{K_{20}}{2(\beta + K_{20})\sigma}y^2\right)$$

$$\hat{\eta}_3 = \gamma_2 + \frac{K_{20}}{2(\beta + K_{20})\sigma}y^2. \tag{10.13}$$

Thus, system (10.7) acts as the master system, and systems (10.9) and (10.12) form the slave system. For the synchronization problem, the unknown states are obtained with (10.10) and (10.13). For the anti-synchronization problem, the same equations are used, but defining the states $\hat{\xi}_2 = -\hat{\eta}_2$ and $\hat{\xi}_3 = -\hat{\eta}_3$.

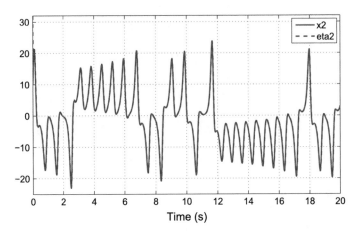

Fig. 10.1 Synchronization between x_2 and $\hat{\eta}_2$

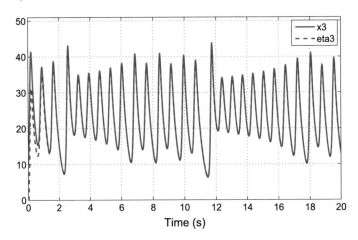

Fig. 10.2 Synchronization between x_3 and $\hat{\eta}_3$

Simulations were performed during 20 s, with $\alpha = 0.993$, parameters $\sigma = 10$, $\rho = 28$, $\beta = 8/3$ and gains $K_{10} = 250$, $K_{11} = 1$, $K_{20} = 150$ and $K_{21} = 0.01$. The initial conditions were set as $x_1(0) = x_2(0) = x_3(0) = 10$, $\gamma_2(0) = \gamma_3(0) = -10$.

Figures 10.1 and 10.2 show the synchronization between the states x_2 and x_3 of the master and their estimations $\hat{\eta}_2$ and $\hat{\eta}_3$, respectively, from the slave. Figures 10.3 and 10.4 show the anti-synchronization between the states x_2 and x_3 of the master and their estimations $\hat{\xi}_2$ and $\hat{\xi}_3$, respectively, from the slave. Finally, Fig. 10.5 shows the synchronization and anti-synchronization signals in state space.

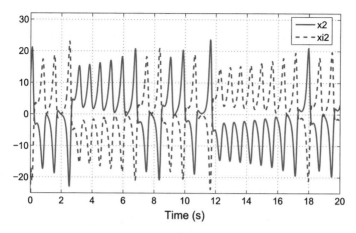

Fig. 10.3 Anti-synchronization between x_2 and $\hat{\xi}_2$

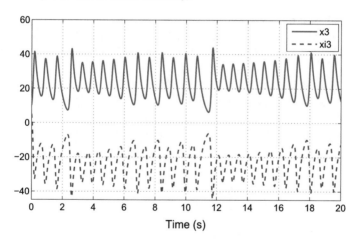

Fig. 10.4 Anti-synchronization between x_3 and $\hat{\xi}_3$

10.2.2   Fractional Rössler System

This attractor was designed Otto Rössler in 1976; its original theoretical equations were later found to be useful in modeling equilibrium in chemical reactions [22]. The generalization to fractional dynamics was proposed in [23].

Consider the incommensurate fractional order chaotic Rössler system:

$$D^{\alpha_1} x_1 = -x_2 - x_3 \tag{10.14}$$
$$D^{\alpha_2} x_2 = x_1 + a x_2$$
$$D^{\alpha_3} x_3 = 0.2 + x_3(x_1 - 10)$$
$$y = x_2,$$

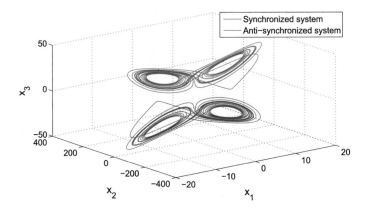

Fig. 10.5 Synchronization and anti-synchronization of the Lorenz system in state space

where a is allowed to be varied.

First, it has to be verified that the unknown states, x_1 and x_3, satisfy the IFAO condition. From (10.14), the following equations can be obtained:

$$x_1 = \phi(y, D^{\alpha_2} y) = -ay + D^{\alpha_2} y \qquad (10.15)$$

$$x_3 = \phi(y, D^{\alpha_1} y, D^{\alpha_1 + \alpha_2} y) = -y + aD^{\alpha_1} y - D^{\alpha_1 + \alpha_2} y, \qquad (10.16)$$

and thus, both states are algebraically observable, so the ROFIO can be built.

Let $\eta_1 = x_1$. Using (10.15), the ROFIO for this variable is

$$
\begin{aligned}
D^{\alpha_2} \hat{\eta}_1 &= K_{10}(\eta_1 - \hat{\eta}_1) + K_{11} I^{\alpha_2} (\eta_1 - \hat{\eta}_1) \\
&= K_{10} \left(D^{\alpha_2} y - ay - \hat{\eta}_1 \right) + K_{11} I^{\alpha_2} (D^{\alpha_2} y - ay - \hat{\eta}_1).
\end{aligned}
$$

Define the auxiliary variable $\gamma_1 = \hat{\eta}_1 - K_{10} y$. Then,

$$
\begin{aligned}
D^{\alpha_2} \gamma_1 &= K_{10} \left(D^{\alpha_2} y - ay - \hat{\eta}_1 \right) - K_{10} D^{\alpha_2} y + K_{11} I^{\alpha_2} (D^{\alpha_2} y - ay - \hat{\eta}_1) \\
&= -K_{10} \left(ay + \hat{\eta}_1 \right) + K_{11} y - K_{11} I^{\alpha_2} (ay + \hat{\eta}_1).
\end{aligned}
$$

So, the ROFIO for $\hat{\eta}_1$ is

$$D^{\alpha_2} \gamma_1 = -K_{10} (ay + \gamma_1 + K_{10} y) + K_{11} y - K_{11} I^{\alpha_2} (ay + \gamma_1 + K_{10} y) \qquad (10.17)$$

$$\hat{\eta}_1 = \gamma_1 + K_{10} y. \qquad (10.18)$$

Let $\eta_3 = x_3$. Using (10.16), the ROFIO for this variable is

$$D^{\alpha_2}\hat{\eta}_3 = K_{20}(\eta_3 - \hat{\eta}_3) + K_{21}I^{\alpha_2}(\eta_3 - \hat{\eta}_3) \tag{10.19}$$
$$= K_{20}\left(-y - D^{\alpha_1+\alpha_2}y + aD^{\alpha_1}y - \hat{\eta}_3\right)$$
$$+ K_{21}I^{\alpha_2}\left(-y - D^{\alpha_1+\alpha_2}y + aD^{\alpha_1}y - \hat{\eta}_3\right).$$

Note that this equation deals with both $D^{\alpha_1}y$ and $D^{\alpha_2}y$, so a variable $\zeta = D^{\alpha_1}y$ is defined and is also estimated via ROFIO:

$$D^{\alpha_1}\hat{\zeta} = K_{\zeta 0}(\zeta - \hat{\zeta}) + K_{\zeta 1}I^{\alpha_1}(\zeta - \hat{\zeta})$$
$$= K_{\zeta 0}(D^{\alpha_1}y - \hat{\zeta}) + K_{\zeta 1}I^{\alpha_1}(D^{\alpha_1}y - \hat{\zeta}).$$

Define the auxiliary variable $\gamma_\zeta = \hat{\zeta} - K_{\zeta 0}y$. Then,

$$D^{\alpha_1}\gamma_\zeta = K_{\zeta 0}(D^{\alpha_1}y - \hat{\zeta}) - K_{\zeta 0}D^{\alpha_1}y + K_{\zeta 1}I^{\alpha_1}(D^{\alpha_1}y - \hat{\zeta})$$
$$= -K_{\zeta 0}\hat{\zeta} + K_{\zeta 1}y - K_{\zeta 1}I^{\alpha_1}\hat{\zeta}.$$

So, the ROFIO for $\hat{\zeta}$ is

$$D^{\alpha_1}\gamma_\zeta = -K_{\zeta 0}\left(\gamma_\zeta + K_{\zeta 0}y\right) + K_{\zeta 1}y - K_{\zeta 1}I^{\alpha_1}\left(\gamma_\zeta + K_{\zeta 0}y\right) \tag{10.20}$$
$$\hat{\zeta} = \gamma_\zeta + K_{\zeta 0}y. \tag{10.21}$$

Substitution of (10.21) into (10.19) leads to

$$D^{\alpha_2}\hat{\eta}_3 = K_{20}(-y - D^{\alpha_2}\hat{\zeta} + a\hat{\zeta} - \hat{\eta}_3) + K_{21}I^{\alpha_2}(-y - D^{\alpha_2}\hat{\zeta} + a\hat{\zeta} - \hat{\eta}_3). \tag{10.22}$$

Define the auxiliary variable $\gamma_2 = \hat{\eta}_3 + K_{20}\hat{\zeta}$. Then,

$$D^{\alpha_2}\gamma_2 = K_{20}(-y - D^{\alpha_2}\hat{\zeta} + a\hat{\zeta} - \hat{\eta}_3) + K_{20}D^{\alpha_2}\hat{\zeta}$$
$$+ K_{21}I^{\alpha_2}(-y - D^{\alpha_2}\hat{\zeta} + a\hat{\zeta} - \hat{\eta}_3)$$
$$= K_{20}(-y + a\hat{\zeta} - \hat{\eta}_3) - K_{21}\hat{\zeta} + K_{21}I^{\alpha_2}(-y + a\hat{\zeta} - \hat{\eta}_3).$$

Finally, the ROFIO for $\hat{\eta}_3$ is

$$D^{\alpha_2}\gamma_2 = -K_{20}(y - a\hat{\zeta} + \gamma_2 - K_{20}\hat{\zeta}) - K_{21}\hat{\zeta} \tag{10.23}$$
$$- K_{21}I^{\alpha_2}(y - a\hat{\zeta} + \gamma_2 - K_{20}\hat{\zeta})$$
$$\hat{\eta}_3 = \gamma_2 - K_{20}\hat{\zeta}. \tag{10.24}$$

Thus, system (10.14) acts as the master system, and systems (10.17) and (10.23) form the slave system. For the synchronization problem, the unknown states are obtained with (10.18) and (10.24). For the anti-synchronization problem, the same equations are used, but defining the states $\hat{\xi}_1 = -\hat{\eta}_1$ and $\hat{\xi}_3 = -\hat{\eta}_3$.

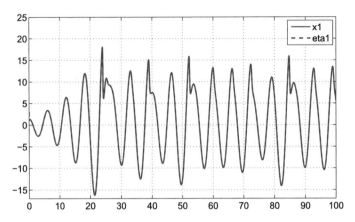

Fig. 10.6 Synchronization between x_1 and $\hat{\eta}_1$

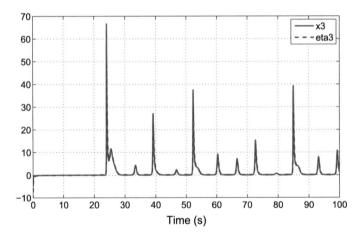

Fig. 10.7 Synchronization between x_3 and $\hat{\eta}_3$

Simulations were performed during 100 s, with $\alpha_1 = 0.9$, $\alpha_2 = 0.8$, $\alpha_3 = 0.7$, parameter $a = 0.63$ and gains $K_{10} = 120$, $K_{11} = 1$, $K_{\zeta 0} = 100$, $K_{\zeta 1} = 1$, $K_{20} = 10$ and $K_{21} = 0.01$. The initial conditions were set as $x_1(0) = 1$, $x_2(0) = 0$, $x_3(0) = -5$, $\gamma_1(0) = 3$, $\gamma_\zeta(0) = \gamma_3(0) = 1$.

Figures 10.6 and 10.7 show the synchronization between the states x_1 and x_3 of the master and their estimations $\hat{\eta}_1$ and $\hat{\eta}_3$, respectively, from the slave. Figures 10.8 and 10.9 show the anti-synchronization between the states x_1 and x_3 of the master and their estimations $\hat{\xi}_1$ and $\hat{\xi}_3$, respectively, from the slave. Finally, Fig. 10.10 shows the synchronization and anti-synchronization signals in state space.

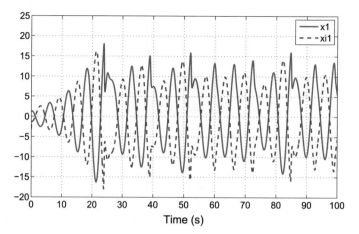

Fig. 10.8 Anti-synchronization between x_1 and $\hat{\xi}_1$

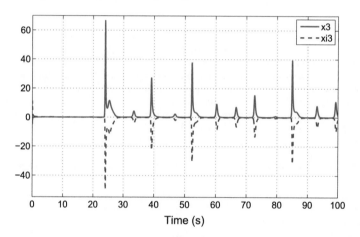

Fig. 10.9 Anti-synchronization between x_3 and $\hat{\xi}_3$

10.3   Concluding Remarks

In this chapter, a reduced-order fractional integral observer was proposed for synchronization and anti-synchronization of commensurate and incommensurate-order fractional chaotic systems. The observer served as a slave system, while the chaotic system acted as the master. To be able to use this observer, the unknown variables of the master systems had to satisfy the fractional algebraic observability condition. The reduced-order fractional integral observer was proven to be Mittag-Leffler stable. Simulations were performed using the proposed methodology, and it was verified that both synchronization and anti-synchronization were carried out successfully for commensurate and incommensurate-order systems.

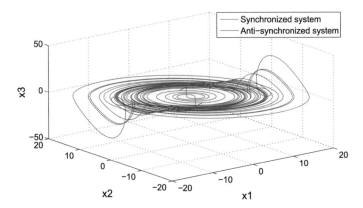

Fig. 10.10 Synchronization and anti-synchronization of the Rössler system in state space

References

1. M. Bennett, M.F. Schatz, H. Rockwood, K. Wiesenfeld, Huygens's clocks, Proceedings: Mathematical, Physical and Engineering Sciences, 458(2019) (2002) 563–579.
2. A. Uchida, Y. Liu, I. Fischer, P. Davis, T. Aida, Chaotic antiphase dynamics and synchronization in multimode semiconductor lasers, Physical Review A 64 (2001) 023801-1–023801-6.
3. I. Wedekind, U. Parlitz, Experimental observation of synchronization and anti-synchronization of chaotic low-frequency-fluctuations in external cavity semiconductor lasers, International Journal of Bifurcation and Chaos 11(4) (2001) 1141–1147.
4. S. Nakata, T. Miyata, N. Ojima, K. Yoshikawa, Self-synchronization in coupled salt-water oscillators, Physica D 115 (1998) 313–320.
5. C.-M. Kim, S. Rim, W.-H. Kye, J.-W. Ryu, Y.-J. Park, Anti-synchronization of chaotic oscillators, Physics Letters A 320 (2003) 39–46.
6. A. A. Emadzadeh, M. Haeri, Anti-Synchronization of two Different Chaotic Systems via Active Control, International Journal of Electrical, Computer, Energetic, Electronic and Communication Engineering 1(6) (2007) 898–901.
7. L. Guo-Hui, Synchronization and anti-synchronization of Colpitts oscillators using active control, Chaos, Solitons & Fractals 26 (2005) 87–93.
8. D. Chen, R. Zhang, X. Ma, S. Liu, Chaotic synchronization and anti-synchronization for a novel class of multiple chaotic systems via a sliding mode control scheme, Nonlinear Dynamics 69 (2012) 35–55.
9. Y. Kawamura, Collective phase dynamics of globally coupled oscillators: Noise-induced antiphase synchronization, Physica D 270(1) (2014) 20–29.
10. J.J. de Espindola, J. Neto, E. Lopes, A generalized fractional derivative approach to viscoelastic material properties measurement. Applied Mathematics and Computation 164 (2005) 493–506.
11. J.J. Rosales, J. F. Gómez, M. Guía, V.I. Tkach, Fractional Electromagnetic Waves, Proceedings of the LFNM*2011 International Conference on Laser & Fiber-Optical Networks Modeling, 4-8 September 2011, Kharkov, Ukraine, p. 1–3.
12. W. Yu, Y. Luo, Y. Pi, Fractional order modeling and control for permanent magnet synchronous motor velocity servo system, Mechatronics 23 (2013) 813–820.
13. J.-D. Gabano, T. Poinot, Fractional modelling and identification of thermal systems, Signal Processing 91 (2011) 531–541.
14. C. Huang, J. Cao, Active control strategy for synchronization and anti-synchronization of a fractional chaotic financial system, Physica A 473 (2017) 262275.

15. A. Razminia, D. Baleanu, Complete synchronization of commensurate fractional order chaotic systems using sliding mode control, Mechatronics 23 (2013) 873–879.
16. A. Razminia, V. J. Majd, D. Baleanu, Chaotic incommensurate fractional order Rössler system: active control and synchronization, Advances in Difference Equations 2011(15) (2011).
17. A. Razminia, D. F. M. Torres, Control of a novel chaotic fractional order system using a state feedback technique, Mechatronics 23 (2013) 755–763.
18. T. Zhou, C. Li, Synchronization in fractional-order differential systems, Physica D 212(1-2) (2005) 111–125.
19. D. Matignon, Stability results for fractional differential equations with applications to control processing, Computational Engineering in Systems Applications (1996) 963–968.
20. E. N. Lorenz, Deterministic Nonperiodic Flow, Journal of the Atmospheric Sciences 20 (1963) 130–141.
21. I. Grigorenko, E. Grigorenko, Chaotic dynamics of the fractional Lorenz system, Physical Review Letters 91(3) (2003) 034101-1–034101-4.
22. O. E. Rössler, An equation for continuous chaos, Physics Letters A 57 (1976) 397–398.
23. C. Li, G. Chen, Chaos and hyperchaos in the fractional-order Rössler equations, Physica A 341 (2004) 55–61.

Appendix A
Integer-Order System

A.1 Static-State Feedback Control

To illustrate the results of the static-state feedback case, we consider the system defined by

$$\dot{x} = A_0 x + \sum_{i=1}^{n} g_i(x)u + Bu$$
$$y = Cx, \tag{A.1}$$

where

$$A_0 = \begin{pmatrix} -2 & 1 \\ 1 & -2 \end{pmatrix}, \quad g_1(x) = A_1 x = \begin{pmatrix} x_1 x_2 \\ 0 \end{pmatrix} \tag{A.2}$$

$$B = \begin{pmatrix} 1 \\ 1 \end{pmatrix}, \quad C = \begin{pmatrix} 1 & 0 \end{pmatrix} \tag{A.3}$$

Now, we calculate the necessary constant to obtain a numerical bound for the $||x||$, given by (2.38). From $||g_1(x)|| = ||A_1 x|| \leq \mu ||x||^q,$[1] we can choose $\mu = 2$ and $q = 1$. Since A_0 has an eigenvalue with positive real part $\lambda_1 = 1$, we design a control law, $u = Kx$, such that $(A_0 + BK)$ has eigenvalues with strictly negative real part. We choose $K = [-1 \ -1]$, the eigenvalues of $(A_0 + BK)$ remain as $\lambda_{c1} = -1$ and $\lambda_{c2} = -3$. While the eigenvalues of $(A_0 + BK)$ have negative real part, this is possible to find M and ω, such that $||e^{(A+BK)t}|| < Me^{\omega t}, \forall t > 0$. For this example,

---

[1] $||A|| = \sqrt{\lambda_{max}(A * A)}, \ A \in \mathbb{R}^n \times \mathbb{R}^n.$

© Springer Nature Switzerland AG 2018
R. Martínez-Guerra and C. A. Pérez-Pinacho, *Advances in Synchronization of Coupled Fractional Order Systems*, Understanding Complex Systems,
https://doi.org/10.1007/978-3-319-93946-9

the constants are $M = 1$ and $\omega = -1$ that satisfy the inequality. Finally, the interval of the initial conditions can be determined from (2.37), that is

$$||x_0|| \leq \frac{|\omega|}{2\mu M^{q+1}||L||} \leq 1.2 \tag{A.4}$$

Now, it is possible to calculate a numerical bound.

$$||x(t)|| \leq \frac{1.2exp(-t)||x_0||}{(1 - 3.309||x_0||)} \tag{A.5}$$

A.2 Dynamical Control

The original system can be controlled by means of applying a dynamical control u that is to say

$$\dot{x} = Ax + \sum_{i=1}^{n} g(x)_i u_i + Bu \tag{A.6}$$

The results obtained for the exponential stability are illustrated with a particular example of the form (A.6). The transformed systems for (A.1) is given from the change of variable $y = x_1 = z_1$, the canonical form obtained is

$$\begin{aligned}
\dot{z}_1 &= z_2 \\
\dot{z}_2 &= \mathscr{F}(z_1, z_2, u) + \dot{u} \\
\dot{u} &= -\mathscr{F}(z_1, z_2, u) + kz
\end{aligned} \tag{A.7}$$

The values for the constants are calculated analogously to the static-state feedback case, and the values are $\mu = 2, q = 2, \rho = 2, M = 1$, and $\omega = -3$. Substituting these values in (2.44) and in inequality (2.46), we get a function that is a bound of the norm of the solution $||x||$.

$$||x(t)|| \leq \frac{x_0 exp(-3t)0.53}{(1 - v(\varepsilon_0))^{1/5}}$$

with the initial condition that satisfies $||x_0|| \leq \varepsilon_0 = 0.53$ (Fig. A.1).

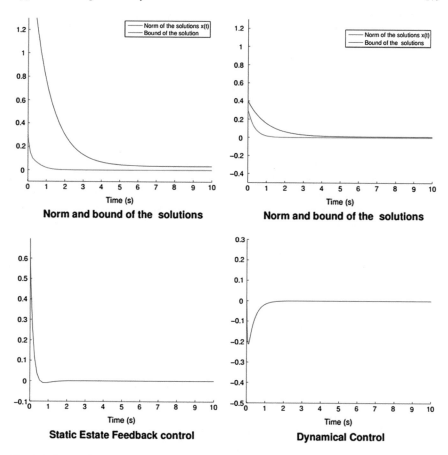

Fig. A.1 Integer Systems

Appendix B
Fractional-Order System

B.1 Static-State Feedback Control

Let the fractional input-affine system, given by

$$D^\alpha x = A_0 x + A_1 x u + Bu$$
$$x(0) = x_0 \quad 0 < \alpha < 1 \tag{B.1}$$

with

$$A_0 = \begin{bmatrix} 0 & 1 \\ 1 & -1 \end{bmatrix}, \quad A_1 x = \begin{bmatrix} 1 & 0 \\ 2 & -1 \end{bmatrix} \begin{bmatrix} x_1 \\ x_2 \end{bmatrix}, \quad B = \begin{bmatrix} 1 \\ 1 \end{bmatrix} \tag{B.2}$$

To calculate the bound of the norm of $x(t)$ for the system (B.1) it is necessary to calculate the constants given in the previous results, first given a gain $K = [-1, -1]$, the matrix $W = A_0 + BL$ satisfies the condition $|arg(\lambda(W))| > \alpha \frac{\pi}{2}$ also the inequality $||A_1(x(t))|| \le \mu ||x(t)||^q$ is satisfied for the values $\mu = 3$ and $q = 1$. On the other hand $||W|| = 2$. From the Corollary 2.1 is obtained $\theta = 1$.

Now, given the constants above it is possible to find and interval for the initial condition

$$||x_0|| < \left(\frac{\alpha ||W||}{2\mu ||L|| \theta^{q+1}} \right)^{\frac{1}{q}} < 0.14 \tag{B.3}$$

as well as, we have

$$||x(t)|| \le \frac{\frac{||x_0||}{1 + 2t^{(0.5)}}}{\left(1 - 50.91 ||x_0|| \left(1 - \frac{1}{1 + ||2||(\frac{t}{2})} \right) \right)} \tag{B.4}$$

© Springer Nature Switzerland AG 2018
R. Martínez-Guerra and C. A. Pérez-Pinacho, *Advances in Synchronization of Coupled Fractional Order Systems*, Understanding Complex Systems, https://doi.org/10.1007/978-3-319-93946-9

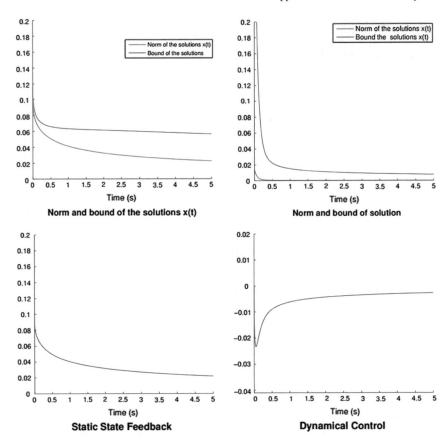

Fig. B.1 Fractional Systems

B.2 Dynamical Control

The fractional transformed system obtained for (B.1) is given by

$$
\begin{aligned}
D^\alpha z_1 &= z_2 \\
D^\alpha z_2 &= \mathscr{F}(z_1, z_2, u) + D^\alpha u \\
D^\alpha u &= -\mathscr{F}(z_1, z_2, u) + kz,
\end{aligned}
\tag{B.5}
$$

where the values constants are calculated analogously to the integer case, and the resulting values are $\mu = 2$, $q = 2$, $p = 4$, and (2.65), we get a function that is a bound of $||x||$, with the initial condition that satisfies $||x_0|| \leq \varepsilon_0 = 0.2$ as well as,

we get $||x(t)|| \leq \dfrac{\dfrac{\varepsilon_0}{1 + 2t^\alpha}}{v(t, \varepsilon_0)^{1/(8)}}$ (Fig. B.1).

Index

© Springer Nature Switzerland AG 2018
R. Martínez-Guerra and C. A. Pérez-Pinacho, *Advances in Synchronization
of Coupled Fractional Order Systems*, Understanding Complex Systems,
https://doi.org/10.1007/978-3-319-93946-9

Printed in the United States
By Bookmasters